Robotics for Electronics Manufacturing

Understand the design, testing, and application of cleanroom robotics and automation with this practical guide. From the history and evolution of cleanroom robots to the latest applications and industry standards, this book provides the only complete overview of the topic available. Robotics for automating the most demanding cleanroom manufacturing process, the making of semiconductor devices, is used as an example throughout the book. The principles and applications also apply to related industries, including the flat panel display, solar panels, hard disk, nanotechnology, MEMS, and pharmaceutical industries.

With over 20 years of experience in robotics and cleanroom manufacturing, Dr. Karl Mathia covers the relevant subjects for the design and testing of clean robots that operate in both atmospheric and vacuum environments. He provides numerous real-world examples so the reader can learn from professional experience, maximize the design quality, and avoid expensive design pitfalls. The book also provides guidelines and hands-on tips for reducing development time and product cost. Compliance with industry standards for the design, assembly, and handling of cleanroom robots is stressed throughout, and detailed discussions of recommended materials for atmospheric and vacuum robots are included to help shorten product development cycles and avoid expensive material testing.

This book is the perfect practical reference for engineers working with robotics for electronics manufacturing in a range of industries that rely on cleanroom manufacturing.

Dr. Karl Mathia studied in Germany and the United States and holds advanced degrees in Electrical and Computer Engineering. He has over 20 years of experience in research and development, product development engineering, and also held management positions at leading robotic firms, including Brooks Automation and Newport Corporation. Dr. Mathia has published numerous articles in the area of automation, controls, and intelligent systems, and taught short courses in industry. He currently works as Chief Engineer at Zitech Engineering, LLC.

In my 16 years of serving the robotics industry, I have never come across a single book that explains the history, design, and use of cleanroom robotics for electronics manufacturing so thoroughly. The book provides an excellent description of the environment and challenges of this industry and gives valuable insight for designing robots and equipment to meet these challenges. This is a must read for anyone designing cleanroom equipment for electronics manufacturing!

Jeff Baird, Director of Engineering, Adept Technology, Inc.

A must read for anyone working on semiconductor or flat panel robotics. This book captures theory, applications and best practices. Chapters 2, 3, 4 and 7 are a concise reference for designing, specifying and implementing robots. Chapters 5 and 6 provide the technical background to both develop and control robotic systems.

Dr. Martin P. Aalund, Director NPI Engineering, KLA-Tencor Corp.

Karl has created the definitive reference for cleanroom robotics, as well as a practical guide for anyone who wishes to go beyond theory to the economic justifications and real world commercial requirements to deploy robot technology.

Dr. Rich Mahoney, Director of Robotics, Engineering & Systems Division, SRI International

This volume provides a comprehensive view of robot use as part of electronics manufacturing. The book gives a good overview of the different aspects to be considered in the design and deployment of robots for this sector. The text covers a sector overview, in-depth material for different applications areas and discusses also testing and deployment. It is a valuable reference both to engineers and technical managers in the field.

Dr. Henrik I. Christensen, KUKA Chair of Robotics, College of Computing, Georgia Institute of Technology

Robotics for Electronics Manufacturing is an important new reference work for anyone involved with manufacturing robots. The book provides design guidelines for robots in both air and vacuum environments, as well as a thorough overview of robot kinematics and dynamics. The chapter on testing and measuring robot performance is especially valuable as an accessible explanation of the many ISO, ANSI and RIA standards.

Dr. Trevor Blackwell, CEO and Founder, Anybots, Inc.

Robotics for Electronics Manufacturing is a fundamental and thorough reference for engineers practicing, or preparing to practice, automation design for the semiconductor and electronics equipment manufacturing industry. No other reference covers the disparate requirements and best practices for both atmospheric and vacuum robot design, as well as including test and characterization methods which are the key to the successful manufacturing of such products.

Dr. Jeffrey C. Hudgens, Director of Robotics, Applied Materials Inc.

Robotics for Electronics Manufacturing

Principles and Applications in Cleanroom Automation

KARL MATHIA
Zitech Engineering, LLC

CAMBRIDGE
UNIVERSITY PRESS

CAMBRIDGE UNIVERSITY PRESS
Cambridge, New York, Melbourne, Madrid, Cape Town, Singapore,
São Paulo, Delhi, Dubai, Tokyo

Cambridge University Press
The Edinburgh Building, Cambridge CB2 8RU, UK

Published in the United States of America by Cambridge University Press, New York

www.cambridge.org
Information on this title: www.cambridge.org/9780521876520

First published 2010

Printed in the United Kingdom at the University Press, Cambridge

A catalogue record for this publication is available from the British Library

ISBN 978-0-521-87652-0 Hardback

To Maria

Contents

Preface

This book is about the design and application of industrial cleanroom robots in electronics manufacturing. It is intended as a hands-on technical reference for engineers and factory managers involved in manufacturing electronic devices in cleanroom environments. The book provides insight into the principles and applications of industrial cleanroom robotics, in particular in semiconductor manufacturing, the most demanding process in terms of cleanliness requirements. Other examples are the hard disk, flat panel display, and solar industries, which also use high levels of cleanroom automation and robotics. In contrast to the complex manufacturing process, the typical robotic designs often utilize relatively simple robot kinematics in the highly structured environments of process and metrology tools. Some industries, for example the semiconductor front-end industry, are governed by technical standards and guidelines, which are generally helpful during the design process of robotic systems. On the other hand, robotic engineers in electronics manufacturing face challenges that are unknown in other markets, most importantly the cleanliness required in certain factories. Strict cleanliness requirements have resulted in two categories of cleanroom robots: 'atmospheric robots' for high-quality cleanliness at ambient atmospheric pressure, and 'vacuum robots' for extreme cleanliness in enclosures under various vacuum pressures. These two categories are the focus of this book.

The book is organized into seven chapters. Chapters 1 and 2 provide an overview of industrial robotics and industrial cleanroom robotics and are not prerequisites for the technical Chapters 3 to 7:

Chapter 1 provides an overview of the history and different types of industrial robots, and their socioeconomic impact.

Chapter 2 provides an overview of electronics manufacturing in cleanroom environments, cleanliness standards, and the emergence of cleanroom robots in semiconductor manufacturing.

Chapter 3 presents guidelines and best practices for the design of atmospheric robots, including the design example of a wafer-handling robot.

Chapter 4 presents guidelines and best practices for the design of vacuum robots, including the design example of a wafer-handling vacuum robot for automating a six-sided cluster tool.

Chapter 5 reviews common kinematic structures before discussing the kinematics of SCARA-type robots that are commonly used in electronics manufacturing. The forward kinematics model of a three-link robot arm is derived.

Chapter 6 discusses a general dynamic model for robot manipulators and derives the specific model for a three-link robot arm. A decentralized joint control strategy suitable for networked robot control is established.

Chapter 7 introduces several test and characterization methods and their underlying theory. Suitable test fixture designs are described.

A total of 29 examples throughout the book illustrate applications of the presented theory and concepts. All numerical examples were programmed in Matlab®. The International System of Units (SI units) is used whenever possible. For convenience some obsolete units that are still in use are also provided. SI base units, derived SI units, and unit conversion tables for non-SI units are listed in Appendix A. Applicable industry standards are listed at the end of each chapter. Contact information for the relevant publishing standards organizations are listed in Appendix B. Standard sets of conditions for temperature and pressure (STP) are listed in Appendix C. These are used to allow comparisons between different sets of experimental data and are relevant for applications in controlled vacuum and atmospheric environments.

Acknowledgments

This book would not have been possible without the advice and support of several engineers and scientists passionate about robotics. Special thanks go to Dr. Martin Aalund, Izya Kremerman, Ken Park, Jeff Thompson, and Enoch Wall. I also thank Andrew Ayre, Robert Bergner, Alan Campbell, Phil Danielson, Walter Henry, William Holtkamp, Dr. Jeff Hudgens, Peter Lundquist, Venu Menon, Frank Pavlik, Dr. John Tenney, and Dalton Victor. NxRev, Inc. kindly provided the Pro/ENGINEER software for creating robot drawings. The advice, guidance, and support from the editorial staff at Cambridge University Press is much appreciated. I am especially grateful to my wife Maria and our children Francisca and Dustin. Without their continuous encouragement and patience this book would have not been possible.

1 Industrial robotics

Robotics refers to the study and use of robots (Nof, 1999). Likewise, industrial robotics refers to the study and use of robots for manufacturing where industrial robots are essential components in an automated manufacturing environment. Similarly, industrial robotics for electronics manufacturing, in particular semiconductor, hard disk, flat panel display (FPD), and solar manufacturing refers to robot technology used for automating typical cleanroom applications. This chapter reviews the evolution of industrial robots and some common robot types, and builds a foundation for Chapter 2, which introduces cleanroom robotics as an engineering discipline within the broader context of industrial robotics.

1.1 History of industrial robotics

Visions and inventions of robots can be traced back to ancient Greece. In about 322 BC the philosopher Aristotle wrote: "If every tool, when ordered, or even of its own accord, could do the work that befits it, then there would be no need either of apprentices for the master workers or of slaves for the lords." Aristotle seems to hint at the comfort such 'tools' could provide to humans. In 1495 Leonardo da Vinci designed a mechanical device that resembled an armored knight, whose internal mechanisms were designed to move the device as if controlled by a real person hidden inside the structure. In medieval times machines like Leonardo's were built for the amusement of affluent audiences. The term 'robot' was introduced centuries later by the Czech writer Karel Capek in his play *R. U. R.* (*Rossum's Universal Robots*), premiered in Prague in 1921. 'Robot' derives from the Czech 'robota,' meaning forced labor, and 'robotnik,' a slave or servant. In *R. U. R.* robots rebel against their human creators and eventually kill them, assuming control of the world. Capek seemed surprised by the enormous interest in his robots: "For myself, I confess that as the author I was much more interested in men than in robots" (Jerz, 2002). Another influential piece of art, Fritz Lang's seminal movie *Metropolis*, was released in 1926. Maria, the female robot in the film, was the first robot to appear on screen.

Isaac Asimov, the ingenious science fiction author, is generally credited with the popularization of the term 'robotics.' He used it in 1941 to describe the study of robots and predicted the rise of a powerful robot industry. The term was first published in his short story 'Runaround' in 1942, and then in 1950 in the collection *I, Robot*, which also introduced his famous Three Laws of Robotics (Asimov, 1950). The Zeroth Law was later added to close some 'loopholes.' The Laws of Robotics and their possible implications for technology attracted significant attention and are a common reference for robotics in the context of artificial intelligence. See, for example, Clarke (1993; 1994).

Denavit and Hartenberg (1955) applied homogeneous transformations for modeling the kinematics of robotic manipulators. The advent of automated flexible manufacturing systems (FMS) in the 1960s established robotics as a scientific discipline. The primary objectives for FMS are reduced labor costs, a high product mix, and a factory utilization near factory capacity. A typical FMS combines industrial robots, an automated warehouse, automated material handling, and complex software systems for simultaneously modeling, operating, and monitoring the factory. Industrial robots are a critical factor in this strategy that minimizes the role of human labor, allowing rapid changes to assembly lines, avoiding costly equipment replacements, and enabling the economical production of customized lots (Aron, 1981; Kahaner, 1991; Megahed, 1993; Sciavicco and Siciliano, 1996). *Industrial Robot*, the first international journal dedicated to robotics, began publication in 1972.

Table 1.1 lists selected milestones in the development of industrial robots, including cleanroom robots in electronics manufacturing (Isom, 2004; Kunii and Port, 2001; Spong and Vidyasagar, 1989). Some emerging areas such as service robots and intelligent robotics are not considered.

1.2 The global robotics industry

In the early 1960s the United States was virtually without competition in robot research and production and led Japan, Europe, and the Soviet Union by several years (Aron, 1981). One of the first industrial robots, the Unimate, was manufactured in the United States in 1961 by Unimation, based on a patent filed in 1954. The Unimate, also called a 'programmable transfer machine,' was designed for material handling. It utilized hydraulic actuators and was programmed in joint coordinates during 'teaching' by a human operator. The angles of the various joints were stored and played back in operation mode. Victor Scheinman's Stanford Arm, an all-electric, six-axis articulated arm designed for tracking arbitrary paths in three-dimensional space, increased the applicability of robots to more sophisticated applications such as assembly and welding. Unimation acquired and further developed the Stanford Arm with support from General Motors, and later commercialized it as the PUMA (programmable universal machine for assembly) model.

The Japanese robot industry was 'jump-started' in 1967 when the Tokyo Machinery Trading Company began importing the Versatran robot from AMF Corporation. In 1968 Kawasaki Heavy Industries entered a technology license agreement with Unimation and in 1969 began to produce robots in Japan. In the late 1970s worldwide interest and investment in industrial robotics increased dramatically, resulting in a remarkable boom in global robot industries. Many start-up companies and several large US and Japanese conglomerates entered the market and began producing industrial robots, some resembling the Unimation models, whose designs were protected by patents in the United States but not in Japan (Nof, 1999).

Table 1.1. Selected milestones in industrial robotics.

Year	Milestone
1954	Devol designs a programmable factory robot (patent granted in 1961) aimed at 'universal automation' (patent granted in 1961). His company was named Unimation.
1956	Devol's design prompts Joseph F. Engelberger to champion industrial robots and make Unimation Inc. the world's robot pioneer.
1959	A prototype Unimate arm from Unimation is installed in a General Motors factory. The first commercial industrial robot is installed in 1961.
1960	AMF Corp. introduces the first industrial robot with a cylindrical coordinate frame, the Versatran by Harry Johnson and Veljko Milenkovic.
1967	Japan imports the first industrial robot, a Versatran from AMF.
1968	Unimation licenses its technology to Kawasaki Heavy Industries Ltd. of Japan. This helps to ignite an explosion of robot development in Japan.
1970	Victor Scheinman at the Stanford Research Institute (SRI) introduces the Stanford Arm, an improvement on the Unimate.
1971	Cincinnati Milacron Inc. markets T3 (The Tomorrow Tool), a computer-controlled robot designed by Richard Hohn.
1973	The Asea Group of Sweden introduces the all-electric IRb 6 and IRb 60 robots designed for automatic grinding operations.
1977	Asea Brown Boveri Ltd. (ABB) introduces microcomputer-controlled robots.
1978	Unimation and GM develop the PUMA (programmable universal machine for assembly) based on Victor Scheinman's robot arm design.
1979	Yamanashi University designs the SCARA (selective compliance arm for robotic assembly). IBM and Sankyo Robotics jointly market this robot.
1979	The semiconductor industry publishes the first standard for 200 mm wafers* (SEMI M1.9-79).
1980	Japan becomes the world's largest robot manufacturer. By 1990, Japan's approximately 40 robot makers dominate the global robot market.
1981	Asada and Kanade build the first direct-drive arm at Carnegie Mellon University.
1984	The industrial robot industry consolidation begins. Most small robot companies go out of business within six years.
1994	The semiconductor industry plans to manufacture devices on 300 mm wafers. The first pilot line is targeted for 1997 and early production is planned for 1998 using a high level of automation.
1995	The second robot boom begins, enabled by the computer power now available. Robot–human interaction is addressed.
1997	Substrate-handling robotic systems begin operation at the first 200 mm wafer fabrication facility, by SGS-Thomson in Catania, Italy.
1997	First publication of standards for 300 mm wafer handling (SEMI M1.15-97).
1999	SEMICONDUCTOR300, a joint venture between Infineon Technologies and Motorola, manufactures the first 64M DRAM on a 300 mm silicon wafer, in Dresden, Germany.
2000	TSMC opens its first 300 mm wafer manufacturing line at a chip foundry in Taiwan.
2012	The semiconductor industry's initial target date to begin manufacturing on 450 mm wafers is 2012.

*A wafer is a semiconductor substrate on which multiple die are fabricated.

During the 1980s robot boom, which automated manufacturing on a large scale, the Japanese industrial robot industry (the number one robot producer since 1980) grew at a faster pace than anyone had estimated. From 1978 to 1990 Japanese industrial robot production grew by a factor of 25. During this rapid growth period the Japan Industrial Robot Association (JIRA) repeatedly corrected its forecasts by +80% and more (Aron,

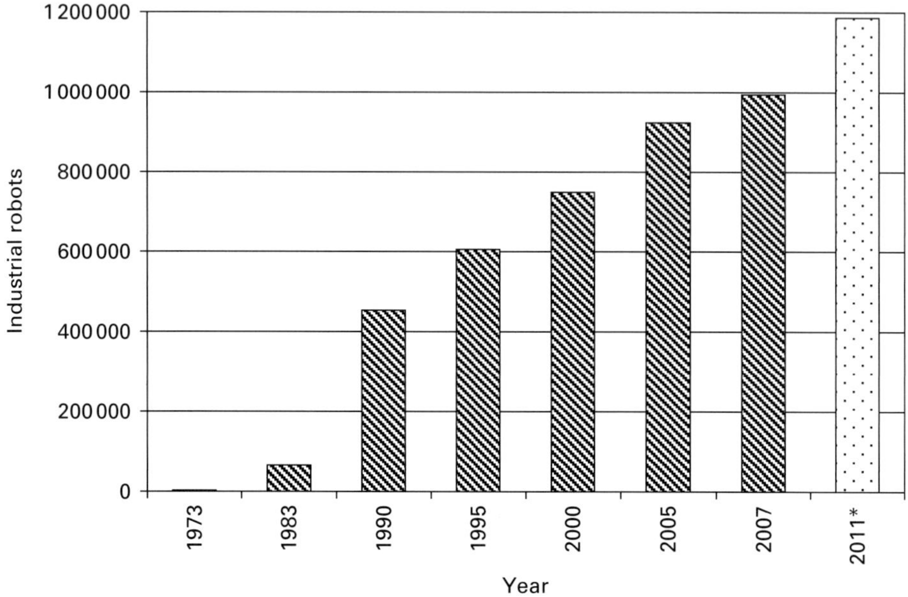

Figure 1.1 Operational stock of industrial robots 1973 to 2011 (* = estimate). Source: World Robotics 2008 (IFR, 2008a).

1981; Kahaner, 1991). (Note: until 2000 Japan used a broader definition for industrial manipulator than the USA and Europe.) The International Federation of Robotics estimates that the worldwide operational stock of industrial robots had reached almost one million in 2007 (Figure 1.1). This number is estimated to increase to almost 1.2 million by 2011. Estimates depend on the assumed average service life, typically between 12 and 15 years, and do not include about 550 000 older robots that had already been decommissioned.

JIRA (now JARA, see Appendix B) attributes this success to three characteristics of industrial robots:

- Industrial robots are programmable automation devices and are, as a consequence, flexible and versatile (unlike special-purpose automated machines).
- Industrial robots exceed the physical and mechanical abilities of humans during extended work periods and in uncomfortable or hazardous environments.
- Industrial robots perform with high fidelity and accuracy and in compliance with their programmed instructions.

Eventually the enormous Japanese robot industry, with its greater financial resources, prevailed in the global competition against its American and European rivals. The first industry consolidation lasted from about 1984, the height of the robot boom, until 1990, and only a small number of non-Japanese companies survived. In 1996, 5 of the 10 largest producers of six-axis robots were Japanese (Schubert, 2005). In 2005 the largest

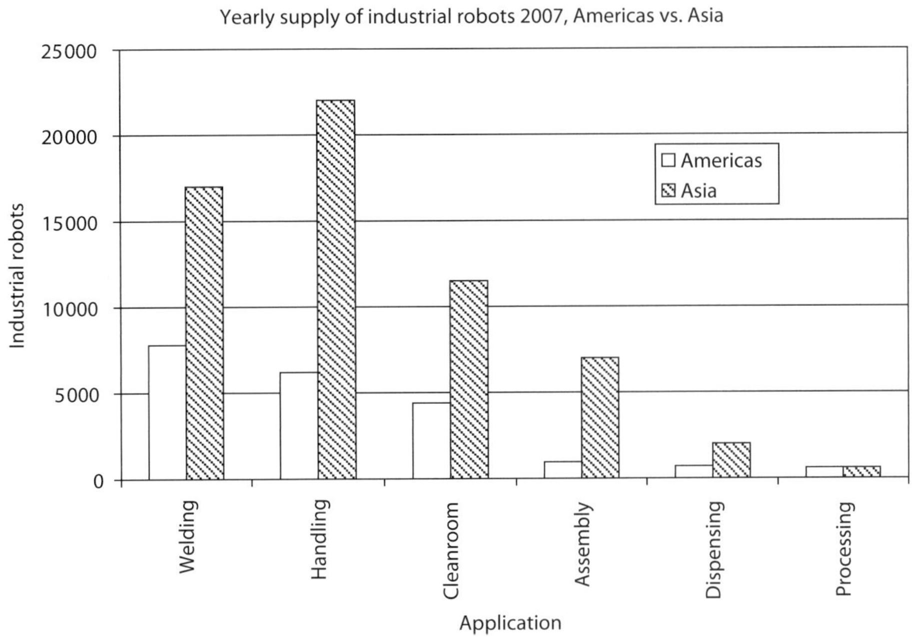

Figure 1.2 Supply of industrial robots by application, Americas versus Asia in 2007. Source: World Robotics 2008 (IFR, 2008b).

industrial robot manufacturers worldwide were Fanuc (Japan), Motoman (Japan[1]), ABB (Sweden), and Kuka (Germany).

1.3 Applications and operational stock by region

The automotive industry was the first to adopt industrial robots on a large scale, primarily for welding applications, followed by the electronics industry with assembly applications. However, the distribution of industrial robot applications varies by economic regions. Figure 1.2 shows the 2007 supply of industrial robots by application for the Americas and for Asia. About 118 000 new industrial robots were supplied worldwide in 2007. The figure shows that in both regions the greatest number of robots was installed for welding and handling, followed by cleanroom applications. The shipment of cleanroom robots increased by a factor of three in the Americas from 2005 to 2007 (IFR, 2008b). In Asia that number decreased in the same time frame.

The global operational stock of multipurpose industrial robots, surveyed by the International Federation of Robotics (IFR, 2007; IFR, 2008b), suggests that Japan's lead position, which the country assumed in 1980, will gradually erode over time. See Figure 1.3. Until the early 1990s installations of multipurpose industrial robots in the European Union (20%) and the United States (7%) reached only a fraction of the number of installations in Japan. Since the mid 1990s the momentum in the industrial robot

[1] Motoman is owned by Yaskawa Electric Corporation, Japan.

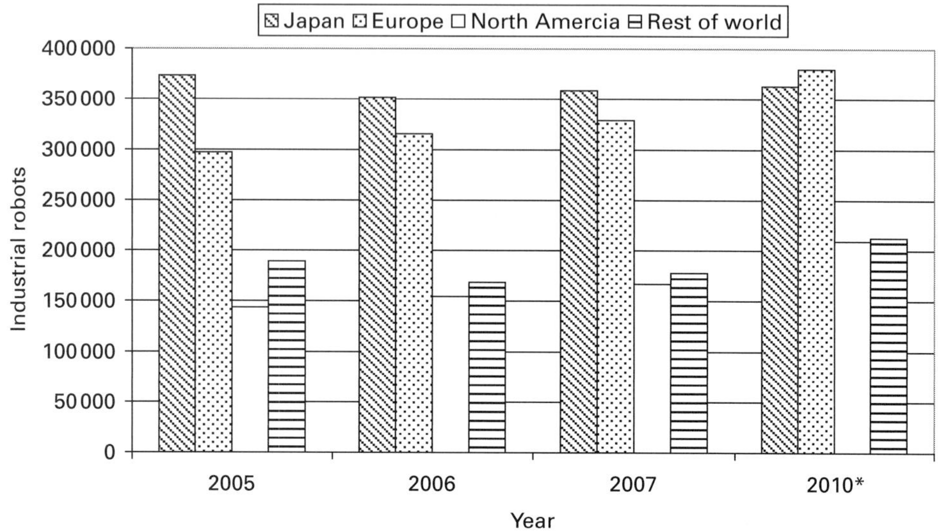

Figure 1.3 Operational stock of industrial robots 2005 to 2010 (* = estimates). Source: World Robotics 2007 (IFR, 2007).

business has shifted to Europe, North America, and other countries. This trend is reported for new installations as well as for operational stock, which has declined in Japan since 1998. Since 2001 the European Union is the leading economic region in installing multipurpose industrial robots.

1.4 Socioeconomic impact

A point of debate about the introduction of robots in America and Western Europe was the loss of jobs caused by robots. This was rarely discussed in Japan, where the positive effects of robots influenced public opinion, in particular the improvement of quality, productivity, and worker safety. It has been argued that the 'labor problem' in the USA and in Western Europe helped Japan to dominate the industrial robot market.

The economic advantage of the industrial robot over human labor is the most important factor in investment decisions, for several reasons (Aron, 1981):

- Reduced labor cost
- Improved productivity
- Improved and more stable product quality
- Resource conservation.

Further reasons are increased workplace safety, increased flexibility of production systems, and labor shortages. The economic advantage is accentuated by the continuously growing gap between increasing labor costs and declining robot costs. Figure 1.4 shows the global price index for industrial robots from 1990 until 2002, established by the United Nations Economic Commission for Europe (UNECE) and the International Federation of Robotics (IFR). The index is adjusted for 1990 currency conversion rates. It

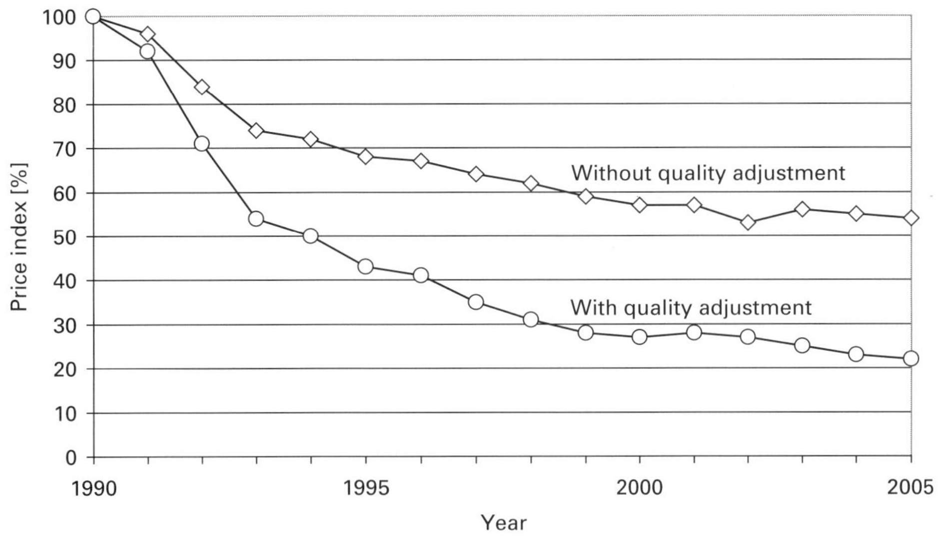

Figure 1.4 Price index for industrial robots 1990 to 2005 (IFR, 2006).

is shown with and without adjustments for quality improvements (IFR, 2006). The price indices were based on the same seven robot models supplied by four major international robot companies with large market shares in Europe and the United States. Without quality adjustment robot prices declined 46% during that time period. Taking into account the enormous performance improvement of robots, the quality-adjusted price decline was 78%.

Robots 'compete' with human workers in manufacturing, at least from an economics perspective, so it seems logical to include the cost of human labor in the price index for industrial robots. The relative prices of industrial robots would fall even more if the (typically increasing) cost of labor compensation is part of the equation. In fact, the IFR study (IFR, 2006) concludes that, with the index of labor compensation included in the price index for industrial robots, robot prices declined by a stunning 90% between 1990 and 2005. Decreasing robot prices, increasing labor costs, and improved robot technology are considered the key drivers for recent, massive robot investment. The amortization period for industrial robots can be as short as one to two years, although the initial cost during the first year after a robot installation can be high, due to production line changes and interest costs. Production slowdowns during robot integration also accrue cost. A case study in Japan found that production initially declined and total costs grew by up to 30%, but at the end of the second year total costs were 25% lower compared to the previous manual production.

1.5 Definitions, standards, and terminology

Standardization is essential for global industries, including not only product standards but also a common terminology and definitions. Cleanroom robotics is a discipline within the broader field of industrial robotics, and is critical, for example, for automating in electronics manufacturing. It is recommended that relevant standards from international standards organizations, or applicable national standards, should be applied. Several definitions for industrial robots, robot systems, and robotics are compared below.

1.5.1 Robot definitions

Japan was the first country to identify robot production as a major strategic industry. Several policies, including standardization, were introduced to popularize robot utilization in manufacturing. The Japanese Electric Machinery Law defined 'industrial robot' in 1971, and additional terminology for industrial robots was standardized in 1979 under the Japanese Industrial Standards (JIS). Other countries followed with national standards, and in 1988 the International Organization for Standardization (ISO) established standards for manipulating industrial robots operated in a manufacturing environment (ISO standard 8373). Most standards and definitions emphasize the flexibility and versatility of 'multipurpose industrial robots.' In contrast, the JIS also included 'dedicated industrial robots' until 2000 (Aron, 1981: Sciavicco and Siciliano, 1996; Megahed, 1993).

Japanese Electric Machinery Law

The Japanese Electric Machinery Law (1971) defines an industrial robot as an all-purpose machine, equipped with a memory device and a terminal device (end-effector), capable of rotation and of replacing human labor by the automatic performance of movements (Aron, 1981).

Japan Robot Association (JARA)

JARA uses a broad classification of industrial robots based on the programming or control method used, formalizing it in the Japanese Industrial Standard JIS B0134:1979, No. 2110–2140. The classes are listed in Table 1.2 (McIntyre, 1997).

American National Standards Institute (ANSI), Robotic Industries Association (RIA)

The ANSI adopted RIA robot standards and defines an industrial robot as "a reprogrammable, multifunctional manipulator designed to move material, parts, tools, or specialized devices through various programmed motions for the performance of a variety of tasks." See, for example, ANSI standard 15.05. This definition corresponds to the JIS classes 3 to 6, but eliminates the manual manipulators and fixed sequence machines.

Association Française de Robotique (AFR)

The Association Française de Robotique (AFR) classifies industrial robots using four types, loosely corresponding to the JIS classes: Type A – telerobotic manipulator; Type B – sequencing robot; Type C – numerically controlled robot; Type D – intelligent robot.

International Organization for Standardization (ISO)

The international standard ISO 8373:1994, originally published in 1988, includes the following definitions related to industrial robots.

Table 1.2. Correspondence between robot definitions published in the Japanese Industrial Standard (JIS) and ISO standards.

Robot category per JIS B0134:1998	Robot definitions per ISO 8373:1994
Sequenced robot (No. 2110)	"Fixed sequence manipulator (No. 2.2): manipulator which performs each step of a given operation according to a predetermined motion operation pattern which cannot be changed without physical alteration." "Sequenced robot (No. 2.10): robot having a control system in which the state of machine movements occurs axis by axis in a desired order, the completion of one movement initiating the next."
Playback robot (No. 2120)	"Playback robot (No. 2.8): robot that can repeat a task program which is entered through teach programming."
Numerically controlled robot (No. 2130)	(ISO standard not available)
Intelligent robot, with one of three control functions (No. 2140):	"Adaptive robot (No. 2.12): robot having sensory control, adaptive control, or learning control functions." (ISO uses the term 'adaptive' instead of 'intelligent'.)
(a) Sensory control (No. 2141)	"Sensory control (No. 5.3.3): control scheme whereby the robot motion or force is adjusted in accordance with outputs of external sensors."
(b) Adaptive control (No. 2142)	"Adaptive control (No. 5.3.4): control scheme whereby the control system parameters are adjusted from conditions detected during the process."
(c) Learning control (No. 2143)	"Learning control (No. 5.3.5): control scheme whereby the experience obtained during previous cycles is automatically used to change control parameters and/or algorithms."
Teleoperated robot (No. 2150)	(ISO standard not available.)

Definition 1.1 (ISO 8373, No. 2.6): "A manipulating industrial robot is an automatically controlled, reprogrammable, multipurpose, manipulator programmable in three or more axes, which may be either fixed in place or mobile for use in industrial automation applications. Note: The robot includes the manipulator (including actuators) and the control system (hardware and software)."

Definition 1.2 (ISO standard 8373, No. 2.14): A robot system is a "system comprising robot, end-effector, any equipment, devices, or sensors required for the robot to perform its task, and any communication interface that is operating and monitoring the robot, equipment, or sensors, as far as these peripheral devices are supervised by the robot control system."

Definition 1.3 (ISO standard 8373, No. 2.15): Robotics is the "practice of designing, building, and applying robots."

The ANSI/RIA standard R15.06 includes a similar definition for industrial robot systems. The above definitions provide the basis for the industrial cleanroom robots presented in Chapter 2.

1.5.2 National and international standards

The International Organization for Standardization (ISO) established robot definitions that correspond to definitions listed in several national standards. Table 1.2 compares the ISO definitions of robot classes with the corresponding robot classes in the Japanese Industrial Standard JIS B0134:1994. The classes also reflect the technical advancement of industrial robots over time: early sequenced robots were numerically controlled manipulators without interaction with the environment through sensors. They were (and still are) programmed off-line. Playback robots are calibrated by 'teaching' through teach pendants, where a human operator guides the manipulator to the desired position, which is recorded for operation. This method is sufficient for 'blind' pick-and-place tasks in structured environments where the locations of tools and work pieces are fixed and well-defined. Intelligent or adaptive robots are equipped with the sensory devices needed for position feedback from the environment, for example proximity sensors or vision systems. This enables feedback control for operating in a changing environment. It also allows the automatic teaching of reference positions (Megahed, 1993; Zhuang and Roth, 1996).

1.5.3 Standard robot types

The main characteristics of industrial robots are

- Number of axes of motion
- Kinematic structure
- Work envelope
- Maximum payload
- Maximum speed
- Accuracy
- Drive train (actuators, remote vs. direct-drive).

Eight robot types are defined in the ISO standard 8373:1994 based on their kinematic structure and the coordinate frame that spans the workspace: Cartesian robots, cylindrical robots, spherical (or polar) robots, pendular robots, articulated (or anthropomorphic) robots, SCARA robots, spine robots, and parallel robots. The preferred kinematic structure of industrial robots depends on the application at hand and is influenced by the required motion, payload, end-effector orientation, and other factors. A robot type typical for electronics manufacturing, the SCARA-type cylindrical robot, is discussed in Chapter 3.

1.6 Applicable and related standards

Several industry standards and guidelines apply directly or indirectly to the robots discussed in this chapter. The following list provides a selection. ANSI standards are published by the American National Standards Institute (ANSI). ISO standards are published by the International Organization for Standardization (ISO). RIA standards are published by the Robotic Industries Association (RIA). SEMI standards are published by Semiconductor Equipment and Materials International (SEMI). Contact information of these organizations is listed in Appendix B.

ANSI/RIA R15.05-1-1990, American National Standard for Industrial Robots and Robot Systems, Point-to-Point and Static Performance Characteristics – Evaluation.

ANSI/RIA R15.05-2-1992, American National Standard for Industrial Robots and Robot Systems, Path-Related and Dynamic Performance Characteristics – Evaluation.

ANSI/RIA R15.06:1999, Industrial Robots and Robot Systems – Safety Requirements.

ISO 8373:1994, Manipulating Industrial Robots – Vocabulary.

SEMI M1.15, Standard for 300 mm Polished Monocrystalline Silicon Wafers (Notched) specifies 300 mm silicon wafer attributes.

SEMI M1.9, Standard for 200 mm Polished Monocrystalline Silicon Wafers (Notched) specifies 200 mm silicon wafer attributes.

SEMI G78-0699, Test Method for Comparing Automated Wafer Probe Systems Utilizing Process-Specific Measurements.

2 Cleanroom robotics

2.1 Manufacturing in cleanroom environments

Clean environments are required for manufacturing modern electronics devices, in particular semiconductor devices, but also hard disks, flat panel displays (FPDs), and solar panels. Wafer processing in the semiconductor industry includes some of the most demanding processes in terms of complexity and cleanliness, due to the submicron dimensions of modern semiconductor devices. This book focuses on industrial cleanroom robotics in semiconductor and FPD manufacturing. Both industries experienced phenomenal technical advancement and growth in the 1980s and 1990s and have established manufacturing facilities in several geographic regions: North America, Europe, and Asia/Pacific Rim. India may emerge as another manufacturing region. The market for semiconductor manufacturing equipment was valued at approximately US$45.5 billion in 2007. The market for FPD manufacturing equipment surpassed the US$1 billion mark in 1997 for the first time. In 2008 it was estimated at US$10 billion.

2.1.1 Cleanroom requirements

Cleanrooms are isolated environments in which humidity, temperature, and particulate contamination are monitored and controlled within specified parameters (SEMI standard E70). Particulates are fine particles, solid or liquid, that are suspended in a gas. Particulate sizes range from less than 10 nm to more than 100 μm. Particulates of less than 100 nm are called ultra-fine particles. Here the term 'particle' is used throughout, representing particles of all applicable sizes, either suspended in a gas or attached to a surface. Cleanroom environments are required if particle contamination is a concern, as is the case, for example, in semiconductor manufacturing. In 1997 the smallest connections in semiconductor devices measured 250 nm. In 2008 that critical dimension was 45 nm, as projected by the International Technology Roadmap for Semiconductors (ITRS). The critical particle size that poses a risk to a device is roughly half of that dimension. It is challenging and expensive to establish and maintain the most stringent levels of cleanroom conditions, especially when human operators, materials, and equipment are constant contamination sources. In fact, one motivation for highly automated fabrication facilities ('fabs') is the removal of humans from the clean manufacturing area. Suitable factory designs and strict protocols are required to maintain cleanroom integrity and minimize the risk of product damage and reduced yields. Typical cleanrooms provide

Table 2.1. Air cleanliness class limits per ISO 14 644–1 and Federal Standard 209E.

ISO 14 644–1 Class	Federal Standard 209E	Contamination limits (particles·m^{-3}) by particle size (μm)					
		≥ 0.1 μm	≥ 0.2 μm	≥ 0.3 μm	≥ 0.5 μm	≥ 1.0 μm	≥ 5.0 μm
1	-	10	2	1	-	-	-
2	-	100	24	10	4	-	-
3	1	1 000	237	102	35	8	-
4	10	10 000	2 370	1 020	352	83	-
5	100	100 000	23 700	10 200	3 520	832	29
6	1 000	1 000 000	237 000	102 000	35 200	8 320	293
7	10 000	-	-	-	352 000	83 200	2 930
8	100 000	-	-	-	3 520 000	832 000	29 300

laminar, vertical flow of filtered air in order to move particles to exit vents at the floor level. Protocols define procedures for operators on how to handle materials, including cleanroom attire and cleanroom-compatible materials for use within a factory (SEMI standards E70 and S20). A high degree of automation and automated materials handling minimizes the involvement of human operators in the manufacturing process. Industrial cleanroom robots became a critical component with the introduction of 200 mm wafers in 1995 and 300 mm wafers in 1998. In fact, robotics is emerging as a production-line integrator for flexible manufacturing strategies, with more than one robot collaborating within a work cell (Defosse, 2004; Kahaner, 1991; Thornton, 2002).

Cleanliness is measured by counting airborne particles under different operating conditions. The International Organization for Standardization (ISO) defines cleanliness classes that specify upper contamination limits per class. The ISO standard replaces the US Federal Standard (FED) 209E, which was officially withdrawn on November 29, 2001. The Japanese robotic industry uses the Japanese Industrial Standard JIS B9920. Table 2.1 lists air cleanliness classes per ISO standard 14 644–1 and the corresponding classes defined in FED 209E (Schicht, 2003). The ISO reference particle diameter of 0.1 μm offers a denomination scheme with simple, single-digit class numbers that correspond with FED 209E. For example, ISO Class 5 corresponds to class 100 in FED 209E.

The measurements performed for cleanroom certification track potential particle sources and their impact on a product. A good practice is first to measure the cleanliness of a cleanroom at the 'as-built stage' without humans and manufacturing equipment present. This step determines whether all filters are functioning to specification and whether there are any leaks. Second, a particle count is performed with the manufacturing equipment under operation. This step isolates the contamination originating from the equipment only. Third, a particle count is performed under normal manufacturing conditions with human operators present. This last step allows an estimate of the impact of humans, their attire, and any manufacturing procedures on the cleanroom environment.

2.1.2 History of cleanroom robotics

Robotics made inroads into cleanroom applications in the 1980s, motivated by increasingly demanding requirements for contamination control, product throughput, and

Table 2.2. Human contamination at different levels of motion.

Human motion	Heat emission (kW)	Moisture emission ($g \cdot h^{-1}$)	Particle emission* (particles$\cdot min^{-1}$)	Breathing requirements ($m^3 \cdot h^{-1}$)
At rest	0.12	90	100 000	0.50
Light work	0.18	180	1 000 000	1.00
4.8 $km \cdot h^{-1}$	0.3	320	5 000 000	2.15
6.4 $km \cdot h^{-1}$	0.4	430	10 000 000	2.55

* Measured particles were 0.3 μm and larger.

product safety. The semiconductor industry began to adapt robot technology from other industries, for example from electronics assembly and automotive manufacturing. The following definition is based on Definition 1.3.

Definition 2.1: Industrial cleanroom robotics refers to the study, design, and use of robot systems for manufacturing in industrial cleanroom environments.

Industrial cleanroom robotics serves three main purposes: (1) reducing costs, (2) improving workplace safety, and (3) improving product yield through cleanliness. Of course, cost reduction and workplace safety are common objectives of automated manufacturing. However, improved product yield through cleanliness is a specific objective in electronics manufacturing. The requirement for extreme cleanliness separates semiconductor manufacturing from other industries. This requires isolating the product from human operators, whose presence makes it difficult to maintain an ultra-clean environment. Table 2.2 lists the measured contamination (particles, moisture, heat) generated by a human at different levels of activity (Egan, 2000). The table emphasizes why contamination from humans is a serious concern: particle emission at rest, even when counting only particles 0.3 μm or larger, exceeds by far the applicable ISO limits in Table 2.1.

2.2 Semiconductor manufacturing

This section reviews the basics of automation in semiconductor manufacturing.

2.2.1 Automation levels and history

Automation in modern semiconductor factories is implemented at three levels:

- Tool automation
- Intrabay automation
- Interbay automation.

Robotics is utilized for material handling at every step within a process or metrology tool, and for material handling between tools. The handling of individual substrates within tools is called tool automation. The transport of substrate carriers between the tools in a process bay (a large cleanroom) is provided by an automated material handling system (AMHS) and is called intrabay automation. The transport of substrate carriers by an AMHS between process bays is called interbay automation.

The first semiconductor factories were extensions of research laboratories, where human operators were vital for process control, substrate handling, and assisting with the frequent equipment failures. Cleanroom robotics was not a pressing issue for several reasons. First, cleanliness was not as critical as in modern factories: device line widths were at 10 μm in the early 1970s and 2 μm in the early 1980s. Second, ergonomics was not an issue: the weight of 50 mm and 100 mm wafer cassettes was suitable for repetitive handling by human operators. Third, cost was less pressing than it is in mature high-technology industries. Fourth, system incompatibilities due to the lack of standards limited the use of robots. Fifth, unreliable semiconductor process tools caused frequent downtimes and therefore limited the benefits of robotics. Finally, suitable robotics technology was not readily available for substrate handling.

Automation and robotics became a priority with the industry transition from 150 mm to 200 mm wafers, and became critical for 300 mm wafer handling in the late 1990s:

- Cleanliness became critical for device line widths below 100 nm.
- Ergonomic concerns prevented humans from repetitively handling fully loaded 300 mm wafer cassettes, a load of about 8 kg.
- Manufacturing costs in the maturing semiconductor industry became a critical economic factor.

Several factors now enable a high degree of automation:

- Industry standards allow the integration of robotic systems with other automation systems and subsystems.
- Modern semiconductor process tools are more reliable, and with longer uptimes automation does improve tool efficiency.
- Suitable cleanroom robotic systems are commercially available.

Figure 2.1 shows the industry transitions from smaller to larger wafer sizes over time. The usual substrates are 'wafers.' Data are shown only for those years in which the number of processed wafers exceeded one million. Transitions occurred roughly every five years. The increase in available wafer area emphasizes the improved economy of scale despite exponentially growing circuit complexity (Grecu and Kücher, 2000). For example, the transition from 200 mm to 300 mm more than doubled the number of dies per wafer, offering a potential 30% cost reduction per die. In fact, if the reduced die size is factored in, the number of devices per wafer increased ten-fold for many logic devices from 1998 to 2008 (Ken Park, personal communication, 2008). The proposed transition to 450 mm wafers in 2012 is also shown. For memory chips the number of bits per wafer increased by a factor of about 1 000 in the same time frame. Automation contributed to this leap in performance. Cleanroom automation and robotics became a crucial factor in the semiconductor business equation during the transition to 300 mm wafers (Aalund and Mathia, 2001; Mathia and Aalund, 2002).

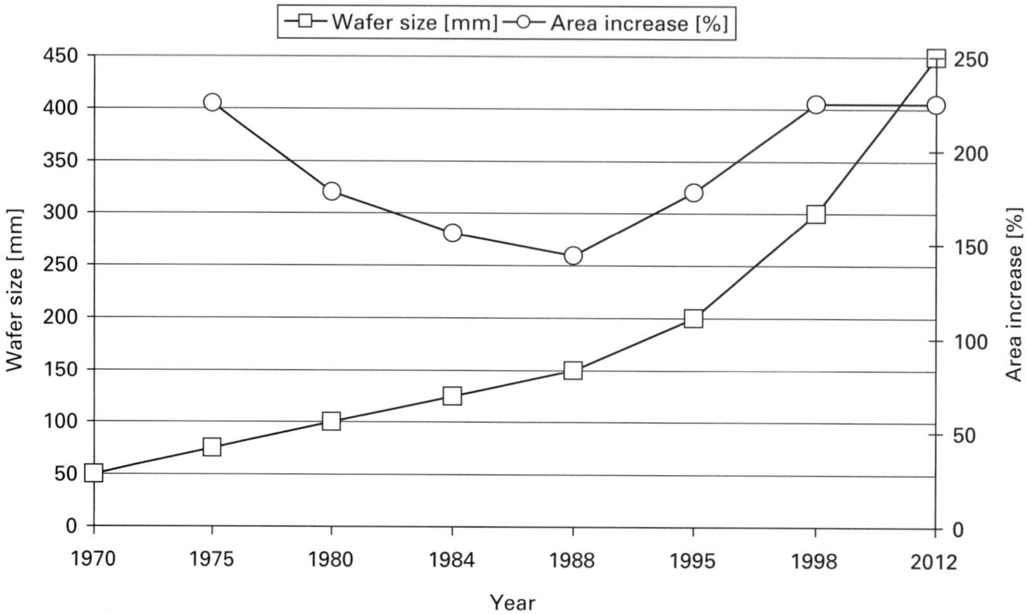

Figure 2.1 Wafer size transition history 1970 to 2012.

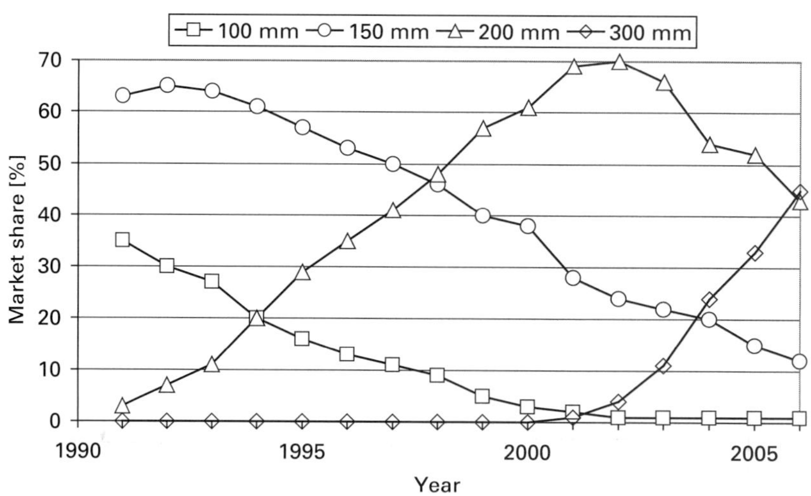

Figure 2.2 Life cycles of wafer sizes from 100 mm to 300 mm, after Vogler (2002).

Figure 2.2 illustrates the life cycles of various wafer sizes from 1991 to 2006 in terms of their market shares. The introduction of a new wafer size requires the industry to build new factories or to 'retool' existing factories. SEMI plans to introduce 450 mm wafers in 2012, although it will take several years to complete the transition, as illustrated in Figure 2.2 for previous transitions. For 300 mm technology a total life cycle of 25 to 30 years is expected.

2.2.2 Semiconductor manufacturing process

The manufacturing or fabrication process for semiconductor devices such as logic and memory includes three process phases: (1) wafer fabrication (producing raw, polished wafers), (2) wafer processing (the 'front-end' manufacturing process), and (3) final manufacturing (the 'back-end' manufacturing process). Figure 2.3 illustrates the front-end process, the most complex of the three process phases. The loops that indicate repetitive process cycles are needed to create several conductive circuitry layers on the device. Dedicated process and metrology tools perform each of the process steps. Inspection and metrology steps are not shown in the figure. The entire process takes several weeks to complete, while wafers travel several kilometers through a factory.

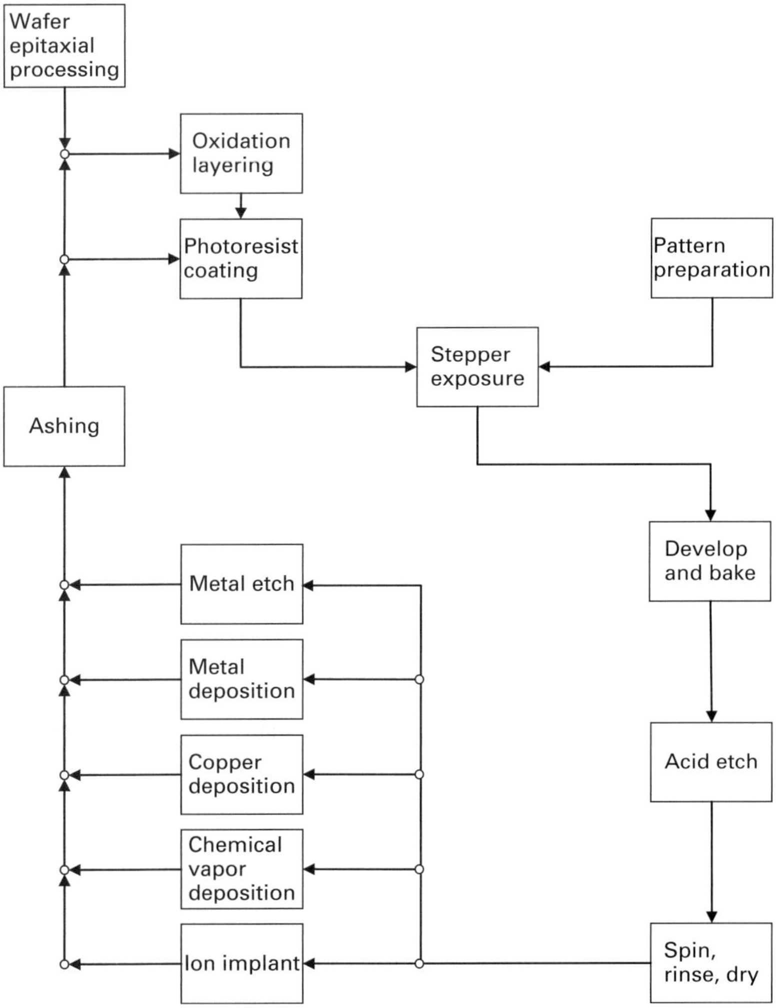

Figure 2.3 Flowchart of a typical front-end semiconductor manufacturing process. Repetitive 'loops' allow the manufacture of multiple circuitry layers.

Dozens or hundreds of process steps are needed, depending on the product. Between the process steps the substrate or wafer is handled by robots or other automation equipment.

The main process steps for the three semiconductor manufacturing phases are summarized below.

Wafer fabrication

- Polysilicon creation: raw polycrystalline silicon is created in a reaction furnace at temperatures exceeding 1000 °C.
- Crystal pulling: silicon crystal ingots are grown by transforming polycrystalline silicon into single crystals with uniformly oriented crystallites.
- Wafer slicing: each silicon crystal ingot is sliced into individual substrates ('wafers').
- Substrate lapping and polishing: the substrate surface is flattened using mechanical lapping and polished using chemical mechanical planarization (CMP).

Wafer processing ('front-end' processing)

- Wafer epitaxial (EPI) processing: a layer of single crystal silicon is grown from vapor onto a silicon substrate at high temperatures.
- Oxidation layering: a thin layer of silicon dioxide or oxide is produced on the substrate.
- Photoresist coating: a uniform layer of photoresist, between 2 and 200 μm thick, is applied to the substrate.
- Pattern preparation: the reticle, the mask with the circuitry pattern for one layer, is placed for the lithography exposure.
- Photolithography (stepper exposure): a device layer is created on the wafer by exposing the photoresist to UV light passing through the pattern mask (reticle).
- Develop and bake: the substrate is developed to remove the exposed photoresist areas. The remaining photoresist is hardened by 'soft-baking.'
- Acid etch: selected areas of material are removed from the substrate using different types of acid, base, or caustic solutions.
- Spin, rinse, dry (SRD): the substrate is repeatedly cleaned to remove any contamination from its surface.
- Ion implant: the electrical characteristics of the substrate layer are changed by bombarding the surface with ions of a particular dopant.
- Chemical vapor deposition (CVD): controlled chemical reactions from various processes create desired device layers on the substrate surface.
- Metal deposition: a conductive layer is created using physical vapor deposition (for aluminum, gold, tungsten) or damascene patterning (for copper).
- Metal etch: conductive circuit paths are created by selectively removing portions of the metal layer. An alternative for many devices is chemical mechanical planarization (CMP).
- Photoresist develop and strip (ashing): the remaining photoresist is removed with an ashing process that uses high-temperature plasma.

Final manufacturing ('back-end' processing)

- Probe test and die cut: the operation of each device on a substrate is tested using automated methods.

- Wafer slicing: the substrate is sliced into individual 'chips.'
- Wire bonding: copper, aluminum, or gold leads are attached to the individual dies via thermal compression or ultrasonic welding.
- Packaging: each device is sealed into a ceramic or plastic enclosure.

2.2.3 Robot applications

Each of the semiconductor manufacturing processes in Figure 2.3 requires a process tool, often with distinctive process requirements that also influence the choice and specifications of the substrate-handling robot. For example, atmospheric and vacuum processes require atmospheric robots and vacuum robots, respectively. These two robot categories are discussed in Chapters 3 and 4.

Atmospheric robot applications

Tools that operate under atmospheric pressure utilize atmospheric robots for wafer handling. Examples are:

- Equipment front end module (EFEM): a standardized mini-environment at the interface between the factory's automated material handling system and one or more process tools (SEMI standard E101-00-1104).
- Chemical mechanical planarization (CMP) tool: removes material from a substrate, creating a very flat surface. CMP allows for more accurate photolithography patterning and film layers with minimal height variations.
- Inspection and metrology tools: needed at various stages of the fabrication process, for example for the detection of contamination defects and for measuring the process quality.
- Rapid thermal processing (RTP) tool: subjects a substrate to rapid temperature bursts from 20 °C to over 1000 °C in less than 10 s. RTP modifies the properties of deposited films.

Vacuum robot applications

Tools that operate under vacuum pressure or in certain (aggressive) gas environments utilize vacuum robots. Examples are:

- Deposition tool: deposition is a fundamental step in semiconductor manufacturing. A layer of dielectric material (insulation) or electrically conductive metal is deposited on the substrate. The main processes are atomic layer deposition (ALD), chemical vapor deposition (CVD), physical vapor deposition (PVD), and epitaxial deposition.
- Etch tool: the substrate is first coated with photoresist, a light-sensitive film. Then a photolithography tool projects the circuit pattern onto the substrate. The etch process selectively removes material from the substrate surface where allowed by the photoresist pattern.
- Ion implementation tool: bombards the substrate with a beam of ions (dopants) that impregnate the film surface of a substrate to the specified depth.

Vacuum robots must withstand various vacuum pressures or (sometimes aggressive) gases or plasma.

2.3 Flat panel display manufacturing

Flat panel display (FPD) manufacturing includes semiconductor technologies and processes. The main differences from semiconductor manufacturing include the vastly larger substrate (panel) area, the relatively large line widths of the circuitry, and the small number of metal layers. FPDs for large television screens utilize liquid crystal technology (O'Halloran et al., 2002). Relevant technology, production steps, and process equipment for liquid crystal displays (LCDs) are outlined below.

2.3.1 FPD market

FPDs are used for computer monitors, televisions, and portable devices. Korea and Taiwan together manufacture about 90% of the world's FPD supply (2008 status). The production efficiency of FPD factories depends on several factors: the substrate size, substrate throughput, manufacturing yield, and product mix. Larger glass substrates ('sheets') offer economy-of-scale advantages by producing more displays from a single substrate. Sheets are categorized by the size of a particular generation. As a result of the lack of industry standards, different substrate sizes are in use. Even the size of a given generation is not always consistent. Typical sheet sizes for each generation are listed in Table 2.3.

Figure 2.4 illustrates the economy of scale of the exponentially increasing sheet area from Generation 1 to Generation 8. FPD factories and their equipment must be regularly upgraded for the next, larger FPD generation. Unfortunately, no standardization body coordinates and encourages technical advancements from generation to generation, as is done by SEMI for the semiconductor industry. A new generation is typically unspecified until some months before factories are planned and equipment is ordered. The robotics design challenges resulting from this fast sheet area increase include:

- The sheet size impacts the robot and end-effector size.
- The sheet weight impacts the robot stiffness and motor power.
- The short life cycle of FPD generations (typically two to three years) requires frequent new designs or design modifications.
- The lack of design standards creates uncertainty and requires accelerated engineering design projects.

Table 2.3. FPD sheet dimensions and area by generation.

Generation	Dimensions (m)	Area (m^2)
1	0.30 × 0.40	0.120
2	0.37 × 0.47	0.174
3	0.55 × 0.65	0.358
4	0.75 × 0.90	0.675
5	1.10 × 1.30	1.430
6	1.50 × 1.85	2.775
7	1.87 × 2.20	4.114
8	2.20 × 2.50	5.500

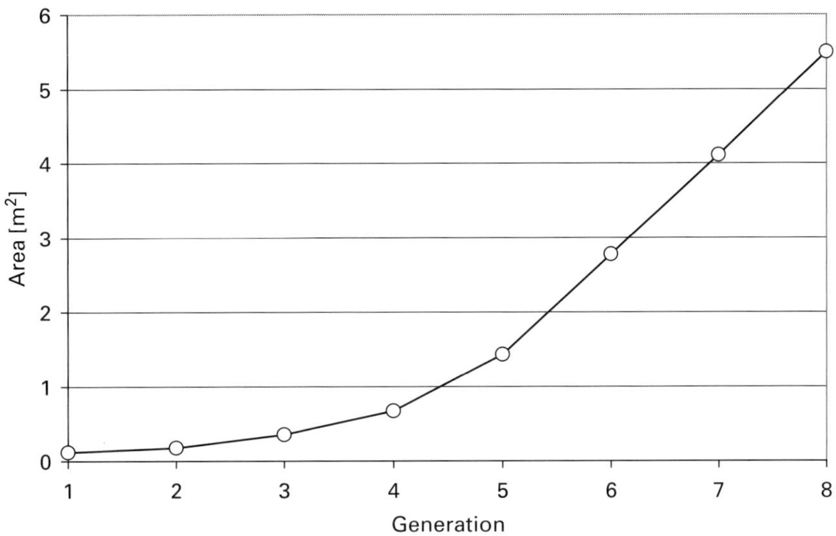

Figure 2.4 FPD sheet area for Generations 1 through 8.

2.3.2 Liquid crystal displays

Liquid crystal displays are made of liquid crystal (LC) compounds that flow like liquids but maintain a short-range crystalline order. The structure of an LC molecule is rod-like, and rotates the direction of polarized light based on its alignment. The molecular alignment, and thus the reflection of polarized light, is controllable by the application of an electric field. The weak intermolecular forces can be overcome, and the LC molecules can be oriented by weak electromagnetic fields. The LC layer in a display is contained between two glass panels. Its thickness is determined by spacers that fix the distance between the glass panels. Seals contain the fluid LC and protect the sensitive circuitry from contamination. Each glass panel has a polarizing layer on the outside. Light passing through one of the polarizers has its polarization rotated to the alignment direction of the liquid crystal. As the light reaches the second polarizer, it can be passed or blocked, depending on the extent of rotation. Transparent conductors on the inner surfaces of the glass plates are used to control the electric field on each cell, and thus the direction of the LC molecules. Depending on the orientation of the molecules, the panel is either transparent or dark. A LC cell acts as a 'light switch.' Light originating from a cold cathode fluorescent lamp or LED in the back of the display can be effectively modulated by using the appropriate conductors in the circuitry. There are passive, active, monochromatic, and color LCDs. In an active matrix display (AMLCD), polysilicon or amorphous thin film transistors (TFTs) are used to activate each pixel. In a passive display, this is accomplished by horizontal and vertical electrodes. There are at least 12 layers in a typical colored AMLCD panel: polarizer, front glass plate (substrate), transparent conductor, passivation layer (or hard coat), polyimide, liquid crystal, polyimide, passivation layer, circuitry, rear transparent conductor, polarizer, and rear glass plate.

2.3.3 FPD manufacturing process

The LCD manufacturing process includes several semiconductor processes, including CVD, lithography, and etch. Fully integrated LCD manufacturing lines include over a dozen individual tools, including (in this order): loader, wet cleaner, polyimide coater, inspection, rubbing, cleaner, after-rubbing cleaner, spacer spray, spacer checker, dispenser, seal dispenser, pre-cure oven, assembly machine, hot press oven, alignment checker, and unloader (robot). Many tools are available as stand-alone machines or are integrated in production clusters. Automated substrate handling is essential. The front and rear glass sheets serve as substrates for several material layers: the front sheet holds the color filter, while the rear sheet is the substrate for the circuitry TFTs.

The front and rear glass sheets are manufactured in different production lines. The two lines merge at an assembly cell, where the glass sheets are combined and sealed. Each pair of glass sheets is large enough to produce several displays. This reduces production costs based on the economy-of-scale principle. Finished panels are cut to fit different products, much like fabrics. The optimal product mix per sheet is flexible and is determined in real time by software tools, depending on any detected defects.

The following summarizes the main steps of a color LCD manufacturing process for TFT technology. Metrology and cleaning steps are not listed.

Front glass manufacturing

- Polarizer is applied to the outside of the front glass sheet.
- Color filter is applied to the inside of the front glass sheet.
- Transparent conductor is deposited using sputtering or printing. A typical material is indium tin oxide (ITO), which serves as the electrode in LCDs.
- A passivation layer, or 'hard coat,' that insulates the ITO electrically is printed, then cured and annealed in a furnace.
- A polyimide (PI) layer is printed using a hot cure process with inert gas. PI rubbing aligns the LC with the polyimide surface parallel to the polarizer direction.
- Spacers, small glass or plastic balls, are sprayed using a dry, semi-dry, or wet process. They ensure a uniform distance between the sheets.
- A seal is deposited and pre-cured in a hot press oven.

Rear glass manufacturing

- Polarizer is applied to the inside (sometimes outside) of the rear glass sheet.
- Transparent conductor is deposited using sputtering or printing.
- Circuitry is created using a series of vacuum processes, including plasma-enhanced chemical vapor depostion (PECVD), sputter deposition, oxidation, lithography, passivation, and etching.
- A passivation layer ('hard coat') is deposited in a furnace.
- A polyimide (PI) layer is printed.

Assembly of the two glass sheets

- The front and rear glass sheets are assembled, including alignment and attachment using UV hardened polymer spots.
- The panel seal is attached using pressure and curing in a clean convection oven.
- The glass sheets are scribed and broken into various display panels.
- The individual displays are filled with liquid crystals using vacuum pressure, and then sealed.
- Polarizer is applied to the front side and the rear side of the LCD panel.
- External contacts are produced by printing with gold or silver paste on the substrate glass.
- The displays are packaged.

Key factors for good process yield are good surface cleanliness, low particle contamination of the manufacturing environment, good gap control, and layer thickness control. Cleanliness requirements are demanding, but not as challenging as in semiconductor manufacturing. For example, the critical particle size for high-resolution, low-power bi-stable displays is 1–3 μm. Particles exceeding that size can cause assembly defects and adversely impact yield. This refers to ISO Class 4 or 5 cleanliness, compared with the usual Class 1 requirements in 300 mm semiconductor fabs.

Organic LED (OLED) technology is becoming available in commercial products and is expected to eventually replace LCD and PDP technology. OLED displays are lighter and offer higher contrast because no backlighting is required.

2.4 Substrate-handling robots

The process flow of the front-end semiconductor manufacturing process in Figure 2.3 indicates that a high level of automation is needed, including substrate-handling robots, to transport and handle the substrates between the many process steps.

2.4.1 Cleanroom technology

Early cleanroom robots

Early cleanroom robots in semiconductor manufacturing replaced technicians who handled wafers with tweezers and vacuum wands. Starting in the 1970s robots were used in process and metrology tools. Fully automated substrate handling emerged in the early 1980s. Typical tool geometries and handling requirements resulted in relatively simple robot kinematics and controller features. However, the demand for tool automation increased with every new wafer size generation, as Moore's Law drove the number of transistors beyond 10 million per cm^2, clock rates beyond 1 GHz, and line widths below 0.1 μm. In particular the introduction of 300 mm wafers and Generation 5 FPDs required advanced automation and robotic systems. By the year 2000 a variety of atmospheric and vacuum cleanroom robots was available for electronics manufacturing in cleanroom environments.

Industry standards and performance specifications

The huge cost and complexity of new 300 mm factories forced the industry to establish standards at the outset to ensure their timely acceptance and implementation. The selection of a robotic system for a given semiconductor or FPD application is influenced by several technical criteria, including wafer throughput, reliability, positional and path repeatability, cleanliness, interoperability with the tool in question, and controller capabilities (Aalund and Mathia, 2001; Manji, 2000). Industry standards and guidelines are available for many criteria. General cleanliness and safety standards are provided by ISO standards (Section 2.1.1). SEMI established industry standards for 300 mm wafer processing, including standards for safety, communication protocols, electrical and mechanical interfaces, and control of the workplace environment in terms of cleanliness and electrostatic discharge (ESD). The throughput of a process or metrology tool is typically defined as the number of processed wafers per hour (wph), which also depends on the time needed for wafer handling and wafer processing. The 'wafer swap time' of a robot is an application-independent throughput metric. It refers to the time needed to replace a processed wafer at a fixed location with an unprocessed wafer, with an assumed zero process time. Selected standards that directly impact automation in the semiconductor front-end industry are listed in Section 3.8.

Robot cleanliness

Cleanliness is an overriding concern in electronics manufacturing, in particular in the semiconductor front-end industry. Cleanliness requires the elimination of contamination sources whenever possible. For example, up to 80% of the yield loss in the production of high-volume, very-large-scale integrated (VLSI) circuits can be attributed to random pattern defects, many caused by contaminating particles that were acceptable at critical dimensions of 1.0 μm, but below 0.25 μm are categorized as 'killer defects' (ICE, 1997). Consequently, ISO Class 1 cleanliness is often required for substrate-handling robots, depending on the application at hand.

Two types of contamination must be addressed during the design process of substrate-handling robots:

- Airborne molecular contamination (AMC) from particles and outgassing generated by the robot
- Contamination of the substrate surface from contact with the robot.

Best design practices, including the selection of suitable materials, minimize particle generation and mitigate the contamination risk. Recommended design guidelines are:

- Minimize the number of moving parts
- Place all moving parts below the substrate
- Enclose and seal the robot interior and evacuate generated particles
- Use internal robot cabling only
- Apply coating or treatment to external robot surfaces
- Use cleanroom approved lubricants
- Use stainless steel screws and washers

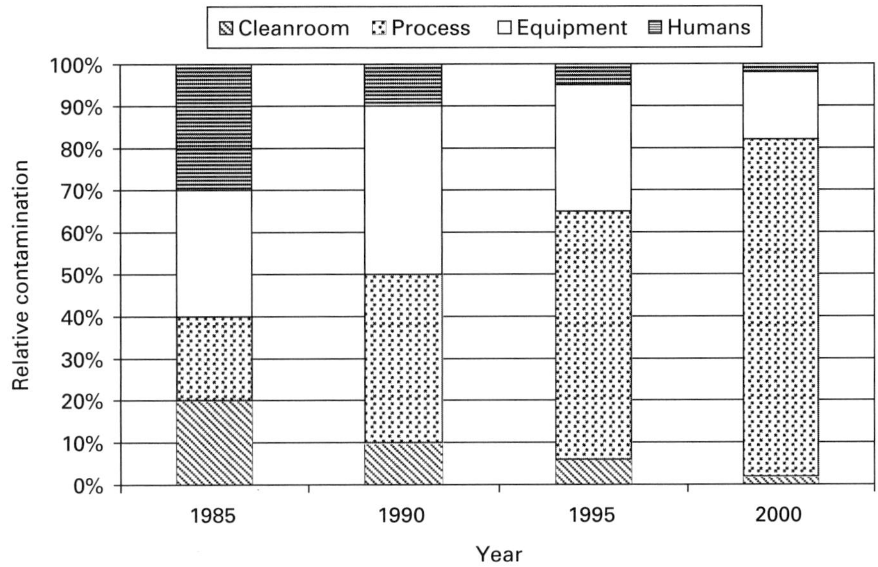

Figure 2.5 Sources of wafer-level contamination, after ICE (1997).

- Minimize the contact surface of end-effectors
- Consider brushless servo motors
- Consider direct-drive trains to eliminate belts and transmissions.

Detailed guidelines and best practices are presented in Chapters 3 and 4. Figure 2.5 demonstrates that the automation strategy of eliminating contamination sources is successful: from 1985 to 2000 the wafer-level contamination from humans and equipment was reduced to less than 3% and 15%, respectively. The 2000 values are estimates. The main contamination source is now the manufacturing process itself.

2.4.2 Economics

Figure 2.6 shows the estimated number of robot shipments to the semiconductor front-end industry (wafer processing). The estimate is based on the number of tool shipments and other secondary indicators (Ken Park, personal communication, 2008). The data shows the industry downturn from 2001 to 2003 and the subsequent recovery. The impact of the 2008–09 financial crisis is not included in the data. The figure also shows that the number of shipped atmospheric robots exceeds the number of shipped vacuum robots by a factor of about three. Possible reasons are:

- The number of shipped process and metrology tools that operate in atmosphere is larger than the number of vacuum tools.
- Many vacuum process tools employ both one vacuum robot (inside the vacuum cluster tool) and an atmospheric robot for tool loading and unloading.

Figure 2.6 also indicates that the average sales price for atmospheric robots continuously decreased from 2000 to 2007. This commoditization trend forced robot companies to

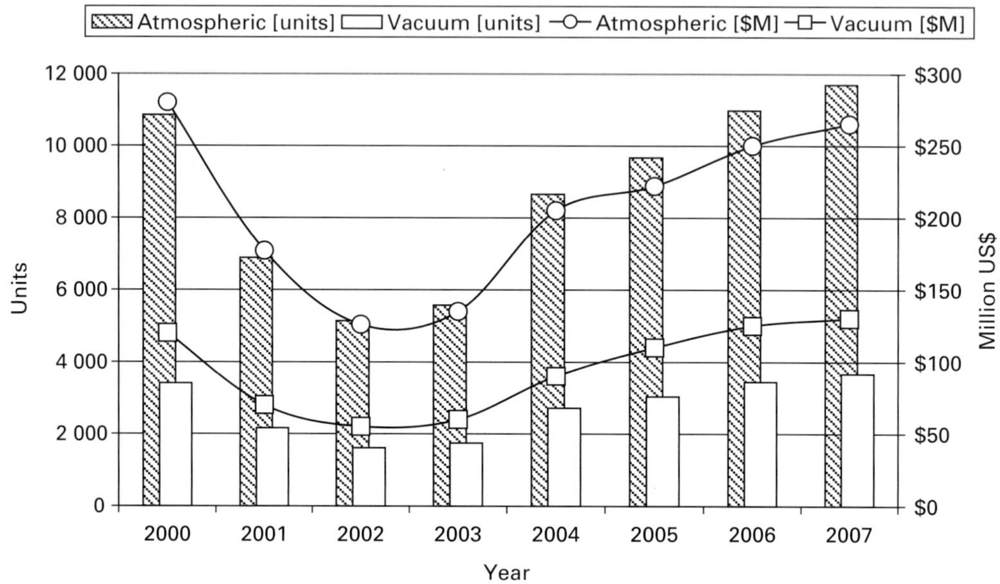

Figure 2.6 Shipment of wafer-handling robots (in units or in millions of US$).

reduce the design and manufacturing cost of atmospheric robots. However, the price of vacuum robots remained fairly constant. This may be due to the small number of suppliers that dominate that market, resulting in an only moderately competitive situation.

2.4.3 SCARA-type robots

Assembly robots

In the 1970s Professor Makino in Japan observed that 80% of assembly movements are vertically within the horizontal reach of a human arm. Motivated by this observation he invented the SCARA robot in 1979. SCARA stands for 'selective compliance arm for robot assembly' (ISO standard 8373:1994, No. 3.15.6); another interpretation is 'selective compliance articulated robot arm'. The term 'compliance' in robotics refers to the elasticity, the inverse of stiffness, of a manipulator, while 'selective compliance' refers to a robot manipulator geometry that provides motion with high rigidity in a selected plane. The kinematic structure of SCARA robots has four axes of motion and is sufficient to move and drop a work piece in any desired position within its workspace. SCARA robots utilize two or three parallel revolute joints to provide compliance in the horizontal plane against vertical loads. When the motors (not the encoders) for the horizontal axes of motion are powered off, the robot can vertically insert a part and compensate for small, horizontal displacements by horizontal sliding into the correct hole or opening.

Handling robots

Substrates in electronics manufacturing are typically handled in horizontal planes at different vertical positions and within the reach of a human arm, often in a cylindrical coordinate frame. This resembles the kinematic structure of SCARA robots for assembly tasks and makes this robot type an obvious choice for substrate handling. There is one exception: the vertical axis of SCARA robots is located at the end of a two-link arm, which is not possible for substrate handling: silicon wafers and other flat substrates require a horizontal, thin end-effector that allows reaching into small openings. Consequently the vertical axis was relocated from the arm tip to the robot centre, creating a 'SCARA-type' robot that became the most common substrate-handling robot in semiconductor and FPD manufacturing. A second reason for moving the vertical axis is cleanliness: it avoids particle generation from moving parts in close proximity to the substrate.

Figure 2.7 shows an example of a SCARA-type robot. Such robots have three or four axes of motion: one for rotating the robot base clockwise and counterclockwise, one for vertical motion, and one for extending and retracting the two-link SCARA arm. The (optional) fourth axis allows horizontal orientation of the end-effector. The arm is attached to the robot base by a revolute joint, sometimes called the 'shoulder.'

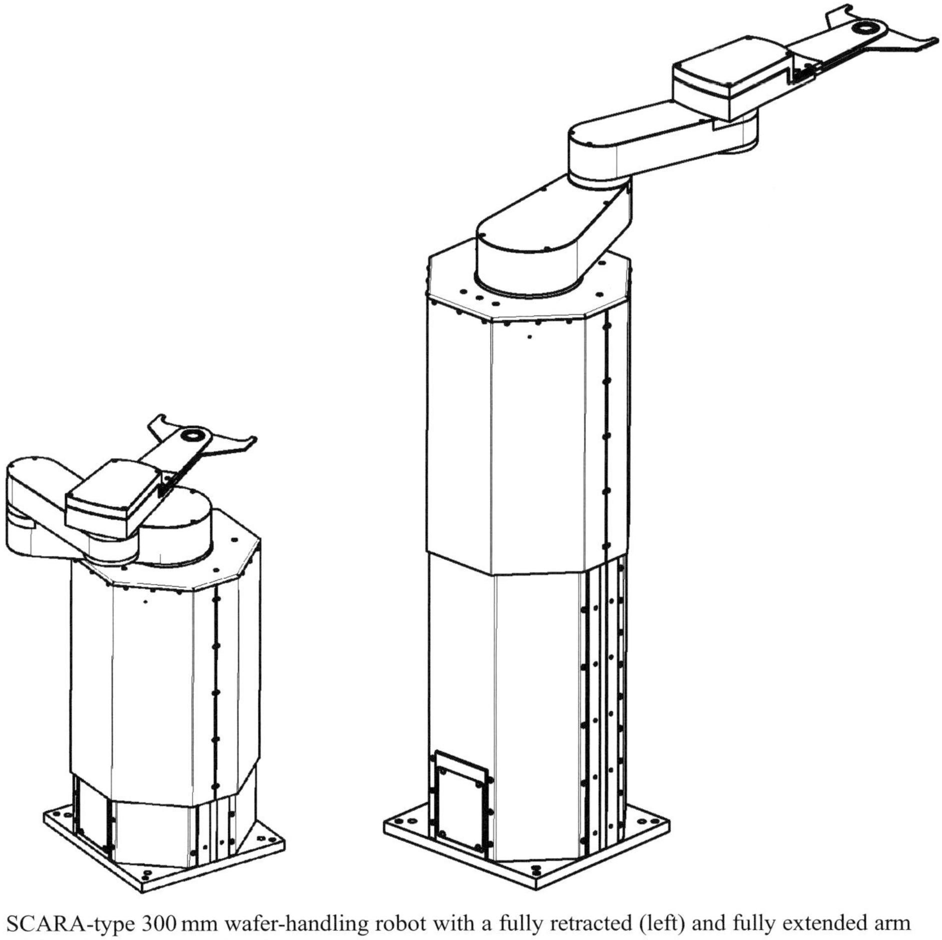

Figure 2.7 SCARA-type 300 mm wafer-handling robot with a fully retracted (left) and fully extended arm (right). Source: Cymechs Corp.

The second revolute joint is the 'elbow,' and the third is the 'wrist.' The vertical axis is usually implemented with a linear ball spline and driven by a ball screw. This SCARA-type kinematic structure has several advantages over more complex robot manipulators:

- Small foot print relative to the work envelope
- Good vertical arm stiffness
- Rugged against minor collisions
- Short pick-and-place time
- Good horizontal repeatability.

The robotic technology used in FPD manufacturing is similar to that in semiconductor manufacturing, although robots are scaled up in size for handling the large glass substrates. For example, the evolution from Generation 5 sheets (1100 mm by 1250 mm) to Generation 7 (1870 mm by 2200 mm) tripled the area and resulted in new requirements for the automation systems, including larger cleanroom robots with heights of up to 3 m. These robots handle FPD cassettes between tools as well as individual panels within a tool (Higuchi et al., 2003).

2.5 Applicable and related standards

Several industry standards and guidelines apply directly or indirectly to the robots discussed in this chapter. The following list provides a selection. ANSI standards are published by the American National Standards Institute (ANSI). ISO standards are published by the International Organization for Standardization (ISO). RIA standards are published by the Robotic Industries Association (RIA). SEMI standards are published by Semiconductor Equipment and Materials International (SEMI). Contact information for these organizations is listed in Appendix B.

ANSI/RIA R15.06:1999, Industrial Robots and Robot Systems – Safety Requirements.

ISO 14 644-1, Cleanrooms and associated controlled environments – Part 1: Classification of air cleanliness.

Sematech ITRS 2008, The International Technology Roadmap for Semiconductors.

SEMI E20-0697, Cluster Tool Module Interface: Electrical Power and Emergency Off Standard.

SEMI E22.1-1296, Cluster Tool Module Interface 300 mm: Transport module end-effector exclusion volume standard.

SEMI E23-1104, Specification for cassette transfer parallel I/O interface.

SEMI E32-0997, Material movement management.

SEMI E70-1103, Guide for Tool Accommodation Process.

SEMI E79-0304, Specification for definition and measurement equipment productivity.

SEMI E101-00-1104, Guide for EFEM functional structure.

SEMI F47-0706, Specification for Semiconductor Processing Equipment Voltage Sag Immunity.

SEMI E20-0697, Cluster Tool Module Interface: Electrical Power and Emergency Off Standard.

SEMI M1.15, Standard for 300 mm Polished Monocrystalline Silicon Wafers (Notched) specifies 300 mm silicon wafer attributes.

SEMI M1.9, Standard for 200 mm Polished Monocrystalline Silicon Wafers (Notched) specifies 200 mm silicon wafer attributes.

3 Design of atmospheric robots

This chapter summarizes engineering guidelines and best practices for the design of cleanroom robots that operate in atmospheric environments. The following definition of atmospheric robots is used here.

Definition 3.1: An atmospheric cleanroom robot operates in cleanroom environments at the ambient atmospheric pressure.

The atmospheric pressure at sea level is 101 325 Pa (1 atm). It decreases exponentially with altitude. A primary design objective for the cleanroom robots discussed here is to contribute to the desired product yield through the damage-free handling of substrates. This can be achieved with a robot design that includes

- Clean materials
- Prevention of electrostatic charge
- Clean drive trains
- Surface finishes
- Robot arm compliance
- End-effectors
- Robot assembly and handling.

Several industry standards provide design guidelines for cleanroom handling systems (Section 3.8). Two contamination sources are not addressed here: (a) outgassing is a concern, but is mainly considered in the design of high-vacuum applications (Section 4.7); (b) cross-contamination can occur when a substrate-handling robot services more than one process chamber.

3.1 Clean materials

Suitable materials must be selected to minimize particle contamination from contact, friction, and outgassing. This section discusses clean materials for parts and components, coatings, the treatment of material surfaces, and lubricants.

3.1.1 Materials for parts and components

Moving or flexing parts are a common source of particle contamination. The following materials are typical candidates for cleanroom robotics based on their cleanliness, durability, and cost: stainless steel, aluminum, plastics, and composites. Non-metallic materials are recommended for surfaces that make contact with substrates.

Stainless steel

Stainless steel is an iron-based alloy that can be plastically formed, for example pounded or rolled. The carbon content ranges between 0.02% and 1.7% by weight. At least 10% chromium and other alloying elements are also used. Stainless steel is a very good material for cleanroom and high-vacuum environments. Its advantages include:

- Excellent cleanliness, even without a special surface finish
- Corrosion resistance
- Very good stiffness (about three times that of aluminum)
- Small thermal expansion (about half that of aluminum).

A disadvantage of stainless steel parts is their high cost compared to aluminum parts. This is the result not only of the material cost but also of the slow (thus expensive) machining that is required for stainless steel. The high density and weight compared to aluminum is a disadvantage for larger moving parts, for example cantilevered robot arms: the dynamics of greater masses requires more energy and an appropriate control algorithm. Therefore stainless steel is used primarily for critical moving parts and smaller parts, for example bearings, lead screws, screws, and washers. It is also used for parts exposed to aggressive environments. A common stainless steel type in cleanrooms is 304L.

Aluminum

Aluminum is a soft and lightweight metal that is widely used when the combined material properties of light weight, strength, and durability are needed. Aluminum and aluminum alloys are typical materials for structural components in cleanroom robotics. For example, aluminum is often the preferred material for larger parts and for components such as robot arms and the robot base. The advantages of aluminum include:

- Good stiffness-to-weight ratio
- Speed and ease of machining
- Cost-effectiveness
- Good cleanliness
- Corrosion resistance and durability
- Non-toxic, non-magnetic, non-sparking
- Availability of surface treatments suitable for aggressive environments.

With only about one-third of the weight of steel, some aluminum alloys have a tensile strength comparable to that of steel. A good stiffness-to-weight ratio enables fast and accurate robot motion while limiting the cost and size of the robot's power train. The cleanliness of aluminum alloy surfaces in dry environments is sufficient for even ultra-high vacuum. Good corrosion resistance and durability is provided by the protective oxide layer that forms on the surface in air. However, in wet environments galvanic corrosion can be rapid, therefore suitable surface finishes are required to prevent corrosion and contamination. Anodization is a surface treatment that creates a hard surface that minimizes particle generation due to contact, friction, and corrosion. Anodized aluminum offers an excellent surface finish for atmospheric applications, but is not suitable for high vacuum. In vacuum the typical protective oxide layer provides sufficient corrosion resistance and durability. Surface finishes are discussed in Section 3.3.

Aluminum alloys are identified by the string format 0000-X0. The first four digits indicate the alloying elements, the letter after the dash identifies the heat treatment, and a one- to four-digit number at the end identifies the specific temperature. For example, 6061-T6 is a free-machining aluminum alloy commonly used in cleanroom robotics. The 6000 category is alloyed with magnesium and silicon. It is easy to machine and can be hardened.

Plastics

Plastics are designed for a specific set of material properties desired for a certain range of applications. In cleanroom robotics a combination of cleanliness and durability is desired, sometimes accompanied by good thermal and chemical resistance. Plastics are traditionally used for manifolds, vacuum lines, O-rings, and gaskets. In cleanrooms a good wear resistance is needed to reduce particle generation from flexing and friction during robot motion. Belts and cables are prominent examples. Outgassing is also a common concern, although primarily in vacuum environments. Nonetheless, plastic parts with low vapor pressure and temperature should be selected whenever possible.

Examples of plastics suitable for cleanroom robotics are:

- PEEK (polyetheretherketone) is a thermoplastic with good chemical and wear resistance, as well as dimensional stability and processing capabilities. See Table 3.1.
- PFA (perfluoroalkoxy) has the elasticity of rubber, good durability, and resistance to oil and solvents. It is used for tubing and similar parts. For example, PFA encapsulated O-rings are listed in the Sematech standard 92051107A-STD.
- PTFE (polytetrafluoroethylene) has good repellent properties and is used in several industries. In cleanroom applications it is used, for example, for bellows, valve manifolds, and pump assemblies.
- Vespel® polyimide has good heat resistance, mechanical stability and durability, and low outgassing. Its low friction coefficient contributes to good cleanliness. Vespel® CP grades are fiber-reinforced polymers with good strength-to-weight ratios.
- Viton® is a fluoroelastomer, a fluorocarbon mixed and cured with 'fillers,' acceptors, and curing agents. Viton® fluoroelastomers are categorized under the ASTM D1418 and ISO 1629 standards as 'FKM.' These plastics offer several good properties for cleanroom use.
- Silicone rubber is a polymer with an operating temperature range from –100 °C to +500 °C. Its tensile strength, elongation, and tear strength are suitable for extreme conditions.

Note that PVC is not listed above. It is not recommended for cleanrooms because it includes several materials prone to outgassing.

Table 3.1. Physical properties of pure PEEK

Mechanical	Thermal	Electrical
Density: $1310\,kg \cdot m^{-3}$ at 23 °C (296 K) (ASTM D792)*	Melting point: 340 °C (613 K)	Dielectric constant: 3.2 at 60 Hz and 23 °C (ASTM D 150, IEC 60250)
Tensile strength: $9.6 - 10.0 \cdot 10^7$ Pa Tensile modulus: $3.5 \cdot 10^9$ Pa at 23 °C (ASTM D638)	Maximum continuous operating temperature: 250 °C (523 K)	Dissipation factor: 0.001 – 0.004 at 60 Hz and 23 °C (ASTM D 150, IEC 60250)
Flexural strength: $1.7 \cdot 10^8$ Pa Flexural modulus: $4.1 \cdot 10^9$ Pa at 23 °C (ASTM D790)	Thermal conductivity: $0.25\,W{\cdot}K^{-1}{\cdot}m^{-1}$	Volume resistivity: $10^{-17}\,\Omega{\cdot}m$ (ASTM D 257, IEC 60093)
Tensile elongation at break: 25% (ASTM D638)	Thermal expansion coefficient: $5 \cdot 10^{-5}\,K^{-1}$ at 416 – 533 K (143 – 260 °C) (ISO 11359, ASTM D 696)	Surface resistivity: $10^{-16}\,\Omega$ (IEC 60093)
Impact strength, notched: $50\,J{\cdot}m^{-1}$ at 23 °C (ASTM D 256)	Flammability (UL 94): V-0	–
Coefficient of friction: 0.2 – 0.3 at $5 \cdot 10^4$ Pa and $0.6\,m{\cdot}s^{-1}$	–	–

* The standards referenced in Table 3.1 are listed in Section 3.8.

These plastics have advantages for substrate handling in clean or harsh environments. PEEK is a good example: it is a semi-crystalline polymer with high purity and high strength, and low particle emission. The low friction coefficient of PEEK provides good wear and abrasion resistance and low particle contamination from physical contact. Originally designed for the food industry, PEEK was rediscovered as a suitable material for clean parts that make physical contact with a substrate, for example gripper pads and vacuum chucks. Table 3.1 lists some physical properties of pure PEEK, according to standard test methods published by ASTM International. The data was compiled from several sources, and may vary slightly. Some properties of PEEK enhanced with other materials are listed in Table 3.2.

Several test methods can be used to measure the impact strength of plastics: Izod, Charpy, Gardner, tensile impact, and others. The notched Izod impact strength in ($J{\cdot}m^{-1}$) or ($ft{\cdot}lb{\cdot}in^{-1}$) is common in North America and is specified by ASTM standard D256. It is the ratio of the impact energy of a pendulum to the thickness of the material sample. The test method specified by ISO 180 expresses impact strength in ($kJ{\cdot}m^{-2}$), that is, the ratio of the impact energy to the area under the notch. For pure PEEK an Izod impact strength of about $50\,J{\cdot}m^{-1}$ and $6\,J{\cdot}m^{-2}$ is reported.

Table 3.2. Wear resistance comparison for selected materials (Hoffman, 2003).

Material	Wear rate ($\mu m \cdot h^{-1}$)	Dynamic friction coefficient ($m{\cdot}s^{-1}$)
Plastics (not reinforced)		
PEEK, pure	17.75	0.42
Composites:		
PEEK, carbon-fiber reinforced	2.16	0.29
PEEK, glass-fiber reinforced	2.36	0.26
Vespel® CR-6100	0.69	0.20
PFA, carbon-fiber reinforced	1.19	0.18

PEEK can be reinforced with certain materials if special properties are required: electrical conductivity is achieved by enhancing PEEK with graphite. For example, grounding through conductive end-effectors is needed when electrostatic discharge (ESD) poses a risk of damaging the substrate. The mechanical properties of PEEK can be enhanced with glass fiber and carbon fiber. See Table 3.2. (By definition such a reinforced material is a composite, not a plastic.)

The wear rates of selected cleanroom materials are also compared in Table 3.2. The wear rates were measured at a velocity of $0.13\ m{\cdot}s^{-1}$ using an unlubricated tri-pin on an AISI carbon steel disc with a $4 \cdot 10^{-7}$ m finish under $8.9 \cdot 10^{6}$ Pa pressure.

The particle generation of cleanroom robots from wear can be reduced with the following design features:

- If belts cannot be avoided in the drive train, durable materials like urethane should be used.
- Cable insulation should be selected for durability and cleanroom compatibility, for example low-outgassing fluoride polymer (Nakayama et al., 2005). Silicon rubber may also be used if electrical insulation is needed.
- The uncontrolled flexing and friction of long cables can be reduced with dynamic cable carriers. Cleanroom-compatible cable carriers from plastic materials are available for ISO Class 3 and 4 cleanliness (Pentagon, 2001).
- Integrating controller and robot avoids external signal cables, thus reducing contamination and cost. (An external power cable is needed.)

Composites

High-performance composites are potential alternatives for metal and ceramic components used in cleanroom robotics. A composite bonds two or more materials (as opposed to dissolved materials like alloys). Composites are available in four categories: polymer-matrix composites (PMCs), metal-matrix composites (MMCs), carbon-matrix composites (CAMCs), and ceramic-matrix composites (CMCs). Their general properties are high strength, high stiffness, low density, strong fatigue and creep resistance, low thermal expansion coefficient, and good chemical resistance. For robot handling the desired high strength, high stiffness, and low density can be provided by fiber-reinforced epoxy (a PMC), aluminum–silicon carbide (AlSiC) composites, or aluminum containing silicon

Table 3.3. Comparison of composites, metals, and a ceramic (Hoffman, 2008).

Material	Density (kg·m^{-3})	Flexural modulus (10^9 Pa)	Thermal expansion coefficient (10^{-6} K^{-1})
Stainless steel (25 °C)	7750–8100	190–210	9–20
Aluminum	2600–2800	70–79	21–25
Alumina (Al_2O_3)	3970	330	8.1
CRE composite	1500–1800	303	1–3
AlSiC composite	2900–3000	300	4–16

Table 3.4. Stiffness and weight data of 300 mm end-effectors, after Hoffman (2003).

End-effector property	Aluminum	Alumina	CP-9800
Weight (kg)	0.812	1.202	0.499
Deflection, no load (mm)	6.096	1.524	0.762
Deflection, from load (mm)	1.524	0.254	0.254
Deflection, total (mm)	7.366	1.778	1.016

carbide particles (a MMC). Carbon fibers are a common PMC reinforcement, with an elastic modulus ranging from 235 to 895 GPa, and a tensile strength ranging from 3200 to 7000 MPa (Hoffman, 2008). Table 3.3 compares properties of two composites, carbon-fiber-reinforced epoxy (CRE) and aluminum–silicon carbide (AlSiC), with those of two metals and one ceramic that are common in cleanroom robotics. Note that, despite their relatively low densities, the composites have a better flexural modulus and a smaller thermal expansion coefficient. A barrier for wide acceptance of composites in industrial cleanroom robotics is their relatively high cost.

A robotics application for composites in electronics manufacturing is their use in end-effectors. Blade and fork-type end-effectors are long and flat for silicon wafers and other substrates. An end-effector made of the CRE composite Vespel® CP-9800 was compared with end-effectors with the same shape and thickness (3 mm), made of aluminum and of the ceramic alumina (aluminum oxide, Al_2O_3). The vertical end-effector deflection with and without the load of a 300 mm wafer was measured. The composite end-effector demonstrated distinct advantages. Its weight was 39% lower than that of the aluminum end-effector, and 58% lower than that of the alumina end-effector (Table 3.4). A reduced manipulator weight is desired for arm compliance, a potential reduction in power requirements, and easier robot control. Table 3.4 also shows a better stiffness-to-weight ratio for the composite end-effector, compared to its aluminum and ceramic equivalents. The CP-9800 end-effector demonstrated the smallest deflection. Furthermore, composites and aluminum are easy to machine, while ceramics must be molded. For example, the integration of wafer-scanning sensors in end-effectors requires embedded wires or fiber optics in the end-effector 'blade.' The channel for the wire is simply machined into a composite or aluminum blade. The good wear rate of composites is evident from Table 3.2. Also, many composites are resistant to chemicals at pH values of 0.2 to 14, which is an advantage for automating aggressive processes.

Table 3.5. Properties of aluminum oxide and silicon carbide, after Accuratus (2009).

Material property	Alumina (99.5% purity)	Silicon carbide
Density (kg·m^{-3})	3 970	3 100
Melting temperature (°C)	2 054	2 730
Maximum operating temperature (°C)	1 750	1 650
Flexural strength (10^9 Pa)	330	550
Tensile modulus (10^6 Pa)	375	410
Thermal expansion coefficient (10^{-6} K^{-1})	8.1	4.0

Many composite materials are available. It is recommended that several manufacturers be consulted and material specifications be compared with the application requirements. Particle tests under realistic conditions may be needed in order to prevent yield problems in the field.

Ceramics

Ceramics are suitable for a variety of substrate handling end-effectors. Various paddle and fork end-effectors for wafer sizes of 100 mm to 300 mm in both atmospheric and vacuum applications are commercially available. However, despite their desirable mechanical, electrical, and thermal properties, only a minority of end-effectors in cleanrooms are made of ceramics. One reason is the relatively high cost: the cost of a ceramic end-effector for handling 300 mm semiconductor wafers can be several times that of the same part made of anodized aluminum. Desirable properties include good cleanliness, stiffness, and durability, and good resistance against wear, heat, and corrosion in aggressive environments.

Ceramics can be coated with protective or smooth materials, for example Teflon®. Unlike metals they are brittle under impact, which would present a serious contamination problem in cleanrooms. Also, ceramics are electrical insulators. If electrostatic discharge (ESD) is a problem, ceramic end-effectors must be grounded to prevent damage to the substrate from ESD events. Common ceramics are alumina (aluminum oxide, Al_2O_3) and silicon carbide (SiC). Alumina has a good stiffness-to-weight ratio and its purity ranges from 99.5% to 99.9%. Silicon carbide is often used for high-temperature and thermal-shock applications. Table 3.5 lists some selected properties of these materials.

3.1.2 Grease and lubricants

Outgassed substances from ordinary grease and lubricants are a contamination risk to substrates, sensors, and instrument optics. Cleanroom-compliant and vacuum-compliant grease and lubricants are chemically stable and non-toxic, synthetic materials with a low vapor pressure, a primary requirement for a low-outgassing material property. Good vacuum lubricants are also 'ultra-filtered' to remove microscopic particles. While suitable greases and lubricants are recommended for atmospheric cleanroom applications, they are critical for high- and ultra-high-vacuum applications.

The three most common lubricants for cleanrooms are hydrocarbon-, silicone-, and PTFE-based (perfluorinated) greases. An important selection criterion is compatibility with a given process or with the materials of the mini-environment, for example with pump fluids and O-rings. A mini-environment is a localized environment created by an enclosure to isolate the product from contamination and people (SEMI standard E63-1104). Vacuum grease can also be used to seal leaking O-rings and other connections. Examples of high-vacuum leak sealants are Vacseal® and Seal-Vac®. The design of vacuum robots is discussed in Chapter 4.

3.2 Prevention of electrostatic charge

Static electricity is an electrostatic charge on a material surface that is caused by an imbalance of electrons. Two possible effects of static electricity are electrostatic discharge and electrostatic attraction.

3.2.1 Electrostatic discharge

An electrostatic discharge (ESD) is defined as a rapid charge transfer between objects of different electrical potentials. It can severely impact the operation of automation and robotic systems and the safety of ESD-sensitive devices. ESD events can occur *to* or *from* an ESD-sensitive device. Here two types of ESD events are of interest: from ESD-sensitive devices to a robot end-effector and from a charged robot to the ESD-sensitive device during a load ('pick') or unload ('place') operation. Note that ESD events can also be caused by a charged human body.

In electronics manufacturing ESD can impact yield and profitability in several ways, including:

- Product damage: catastrophic failure or latent defect, for example damaged devices on a wafer, damaged circuitry on a flat panel display, or damaged magnetic read heads for hard disks, resulting in reduced yield
- Equipment malfunction or failure: damaged robotic system, and tool downtime needed for a system assist, resulting in interruption of the manufacturing process
- Electromagnetic interference (EMI): effects of EMI on electronic devices such as sensors, and signal and communication errors in unshielded cables, resulting in robotic system malfunction and interruption of the manufacturing process.

ESD-sensitive devices include silicon wafers, reticles (photo masks), semiconductor devices, magnetic read heads, thick and thin film resistors, printed circuit boards, and piezoelectric crystals. For example, a silicon wafer, charged with up to 20 kV (in some cases up to 30 kV) during a manufacturing process, can damage the electronics in a robot or robot controller without proper ESD control. Many opportunities for ESD events exist in a manufacturing environment (Kraz, 2003): automated robot handling in electronics manufacturing requires many contacts between the equipment and the product, which increases the probability of triboelectric charging and ESD damage; the shrinking

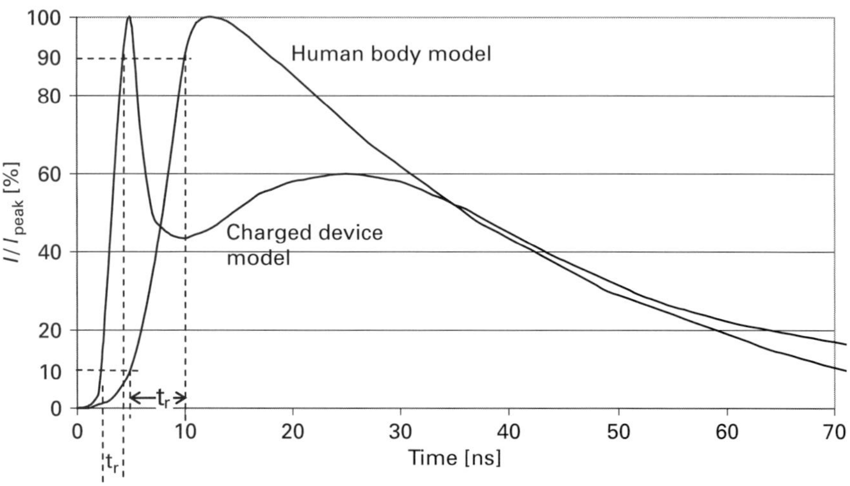

Figure 3.1 Human body model and charged device model, after Björklöf (1999).

geometries of integrated circuits and reticles increase a device's ESD sensitivity and its ability to dissipate the discharged energy or to withstand the voltage levels; magnetic read heads already have a high ESD sensitivity (below 1 V damage level), and are projected to be more sensitive in future products; flat panel displays grow in size while increasing their resolution, which increases their ESD sensitivity.

Three methods are used to model ESD events; see also (ESDA, 2001a) and the ANSI/ESD standard S 20.20 listed in Section 3.8:

- Human body model: simulates discharges from a human body to an ESD-sensitive device
- Machine model: simulates discharges from a robot or tool to an ESD-sensitive device
- Charged device model: simulates discharges from an ESD-sensitive device to a robot or tool.

Standards for these models have been established by the ESD Association: ESD STM5.1-1998 (human body model), ESD STM5.2-1994 (machine model), and ESD DS5.3.1 (charge device model, component level). Figure 3.1 presents an example of the human body model and the charged device model as specified in IEC 61000-4-2. Note that the current discharged during the ESD event is normalized with respect to its peak value (I/I_{peak}). The discharge through a human finger, represented by the human body model, has a rise time of about 5 nanoseconds (ns) from 10% to 90% of the peak value. The discharge from a conductive device such as a silicon wafer or a metal tool has a much shorter rise time. In Figure 3.1 the rise time is about 2.0 ns, but it can be as short as 0.7 ns (Björklöf, 1999). Discharge currents with shorter rise times, higher peaks, and longer pulse widths generally cause more severe damage. The pulse width is an approximate measure of the energy injected into the ESD-sensitive device or equipment.

A field-induced ESD event is a discharge from an ESD-sensitive device that is momentarily grounded through a robot located in an electrostatic field. Field-induced

ESD events can be simulated with the charged device model (CDM). An electrostatic field induces a potential (voltage) between the device and the robot end-effector or arm. If the voltage exceeds the minimum dielectric breakdown strength of air, given the ambient conditions, a rapid energy discharge from the device can occur. The dielectric breakdown strength (the discharge voltage) depends on several factors, including air temperature, pressure, humidity, and electrode shape and size. At standard temperature and pressure (STP) the minimum discharge voltage for air is approximately 380 $V \cdot cm^{-1}$ (Aalund and Mathia, 2007). STP is an internationally accepted set of standard conditions for experimental measurements, intended to enable comparisons between sets of data (Appendix C). The STP conditions for ESD measurements are: an absolute pressure of 100 kPa (1 bar) and a temperature of 273.15 K (0 °C). The SEMI standard E43 provides guidelines for measuring ESD on surfaces.

3.2.2 Electrostatic attraction

Electrostatic attraction (ESA) is a concern in cleanrooms. It is caused by the electrostatic field of a charged surface: the forces created by the field attract contaminating particles of opposite polarity to the surface of wafers, flat panel displays, reticles, and similar products. ESA can be prevented by neutralizing or dissipating the charge. The International Technology Roadmap for Semiconductors (ITRS) warns that the particle adhesion mechanisms at particle dimensions of 500 nm and smaller are not gravity and airflow but mainly forces from electrostatic fields. The critical particle size (diameter) that destroys device functionality was 36 nm in 2006. Unfortunately a charged wafer is difficult to discharge because the processed (front) side must not be touched, and the wafer backside and edge, the only areas where wafers are contacted, is covered with an insulating oxide layer. Dissipative contacts to these areas do not discharge a wafer (Levit et al., 2007). ESA prevention seems to be the best strategy and is preferred over neutralization and dissipation. For example, air ionization can effectively remove surface charges.

The rate at which particles are attracted by an electrostatic field is the particle deposition rate. It depends on several factors, including field strength, field divergence, particle size, and particle charge. An accurate measurement of these parameters is difficult to obtain, so the implementation of good design practices is usually the recommended preventive measure for ESA. The particle deposition rate also depends on the level of airborne molecular contamination (AMC) in the immediate environment of the substrate, and on the time period during which the substrate is exposed to that environment. Three important control variables for minimizing ESA are therefore:

- Electrostatic field strength
- Exposure time
- Level of airborne molecular contamination.

The maximum admissible exposure time is a function of field strength and charge ratio for wafers (M. Aalund, personal communication, 2008). The charge ratio is the ratio of positive to negative ions that are present in the environment. It can be controlled with an ionizer. The admissible exposure time can be increased by decreasing either the charge ratio for a given field strength, or the field strength for a given charge ratio, using a conductive or dissipative surface finish. The exposure time can also be reduced using sophisticated automation and wafer handling. See also Sections 3.2.3 and 3.3.

3.2.3 Controlling the build-up of electrostatic charge

This section discusses methods for preventing static charge build-up on substrate-handling robotic systems, not on the device. The goal of ESD and ESA control is ESD immunity of the substrate-handling robotic system and the capacity to operate without degradation in the presence of ESD events from highly charged substrates. This can be achieved with appropriate engineering design practices and modeled ESD events within specified limits (ESDA, 2001b). ESD events represented by the machine model and the charged device model (contact or field-induced) are considered here. Electrostatic build-up can be prevented with proper grounding and suitable surface finishes. However, environmental ESD control is the primary prevention strategy, typically implemented by the tool manufacturer, not the robot supplier.

Grounding

Effective ESD control cannot be assured without proper grounding of all conductive and dissipative surfaces, a critical component of ESD safety. The standards ANSI/ESD S20.20-2007 and ESD SP10.1 address ESD control (Section 3.8). A robot's resistance to ground, from the end-effector through the entire arm and base, should be measured and reduced if needed. Discharges follow the lowest-impedance path to ground. The ground path can often be rerouted.

A first step in ESD control is to specify the ground path of the electrical current produced by an ESD event. The ground path should bypass sensitive circuitry and electrical parts to avoid potential damage to the robot. Grounding can be improved by reducing the surface resistivity of the parts and components in the ground path. If non-conductive parts are in the ground path, for example end-effector PEEK pads or robot parts made of anodized aluminum, their replacement or treatment will ensure conductivity. For example, anodized aluminum can be replaced with nickel-plated aluminum, and PEEK can be enhanced with graphite. Table 3.6 lists the ranges of the surface and volume resistivity for insulators and for static-dissipative and conductive materials. Gold-plated parts can be used for optimum performance if needed. Grounding tests are performed with ESD guns, which can produce discharges with voltages up to 30 kV at an adjustable

Table 3.6. Typical value ranges for ρ and R_S.

Material type	Surface resistivity R_S ($\Omega \cdot \mathrm{sq}^{-1}$)	Volume resistivity ρ ($\Omega \cdot \mathrm{m}^{-1}$)
Insulators	$> 10^{12}$	$> 10^{9}$
Static-dissipative materials	10^{5} to 10^{12}	10^{2} to 10^{9}
Conductive materials	$< 10^{5}$	$< 10^{2}$

frequency, according to the ESD model used. The user can specify the range for the charge (typically 10^{-6} C) and the pulse duration (10^{-6} to 10^{-9} s). During testing the robot operates under normal operating conditions while being exposed to one or more discharges at the end-effector. This simulates the discharge from a highly charged substrate to a robot end-effector. The robot passes the test if it continues under normal operation as specified by the manufacturer.

Conductive and dissipative surfaces

The use of conductive or dissipative materials and the grounding of all surfaces is a basic equipment design rule. This also applies to robotic systems. Where electrical current may flow across the surface of a given material, the surface resistivity R_S is a convenient entity to estimate the impedance of a ground path, because it can be directly measured using four-terminal (4T) sensing. The measured surface area A_S must be square: the resistance R of a conductor with thickness t, length l, and width w has a cross-sectional area $A = t \cdot w$. Given the (volume) resistivity ρ a well-known relationship can be used (Chase, 2008):

$$R = \rho \frac{l}{A} = \rho \frac{l}{w \cdot t} = \frac{\rho}{t} \frac{l}{w} = R_S \frac{l}{w}, \qquad \text{units}: \ [\frac{\Omega \text{m}}{\text{m}} \frac{\text{m}}{\text{m}}] = [\Omega]. \tag{3.1}$$

With $l = w$ the surface area is square, $A_S = l^2$, and R_S becomes independent of the square size:

$$R = R_S \cdot \frac{l}{w} = R_S, \ \text{units}: \ [\frac{\Omega}{\text{sq}}]. \tag{3.2}$$

This is indicated with the unit 'ohm per square' (Ω/sq). Typical ranges for ρ and R_S are given in Table 3.6 (ESDA, 2001a) for a surface thickness of $t = 1$ mm. R_S of static dissipative materials is between that of conductors and insulators. The standard ASTM-D257 specifies that the surfaces of equipment like robots should have a maximum surface resistivity of 10^6 to 10^9 Ω/sq, which is within the range of static-dissipative materials in Table 3.6. No surface should accumulate a charge that could give rise to electrostatic discharges exceeding 100 V, and isolated surfaces should be grounded through a ground path of approximately 10^9 Ω. Limiting the current resulting from an ESD event also protects personnel.

Environmental ESD control: maximum field strength

Environmental ESD control reduces the electrostatic field strength in the vicinity of the robot and mitigates the risk of ESD events that could damage both the cleanroom robotic system and the handled ESD-sensitive device. For example, Sematech recommends a maximum field strength per 'technology node', the critical-dimension milestones for integrated circuits as promoted by the ITRS. The promoted critical dimension shrinks with each technology node, and the recommended maximum field strength decreases accordingly. Voltages from static charge should not exceed the electrostatic field strength shown for a technology node, for example ±50 V·cm^{-1} for 2009 (Figure 3.2). The recommended limit is projected to be 18 V·cm^{-1} at the 2018 technology node. The

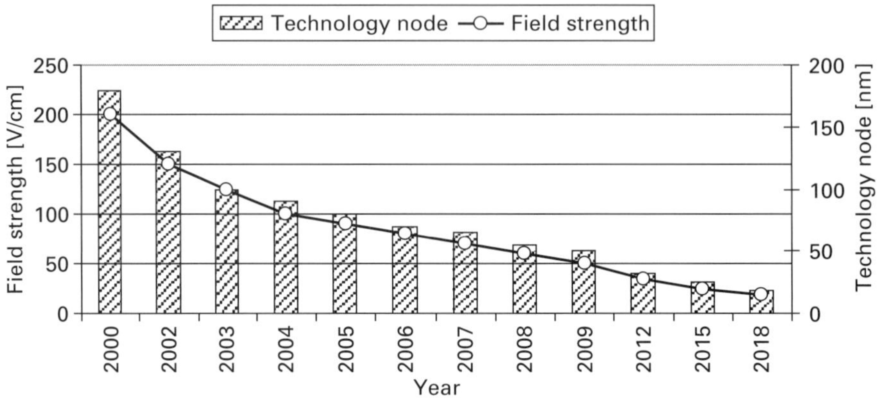

Figure 3.2 Electrostatic field limits per technology node (Aalund and Mathia, 2007).

SEMI standard E129 and ANSI/ESD standard S.20.20 recommend that objects within 0.3 m of an ESD-sensitive product be either static conductive or dissipative, or that potential charges be removed through ionization to prevent ESD events. This includes the surfaces of tools, mini-environments, and substrate-handling robots.

Environmental ESD control: ionizers

Ionization is a key component in maintaining a safe ESD environment. Properly implemented ionizers offer substantial reductions in ESD exposure if properly installed and maintained. Some of the problems observed in the field include poor installation, air blockage, and lack of maintenance (Kraz, 2003).

Environmental ESD control: real-time monitoring

The real-time monitoring of environmental parameters, correlated with historical data, has been proposed for predicting increased risk levels of ESD events. The processing of such data can be used for real-time preventive control of the ESD and ESA risk for both robotic systems and ESD-sensitive devices (Kraz, 2003). A parameter that can indicate an increasing probability of damaging ESD and particle contamination caused by ESA is the static voltage in the production environment. This can be measured through either the voltage on a charged device or through the voltage induced by a device (for example a wafer or flat panel). The following are recommended preventive steps: compliance with ANSI/ESDA standard S.20.20; monitoring of both conductive and static-dissipative grounds; and the wearing by operators of dual wrist straps to ensure a low safe control voltage on all monitored objects. Workstation monitors are available that automatically and simultaneously measure several metal and dissipative grounds, and the proper ground connection of personnel.

3.3 Surface finishes for cleanroom robotics

The interior and exterior surface finishes of cleanroom equipment, including robotic systems, can impact product safety and yield. The surface finishes of cleanroom robots must be inherently clean and protected against corrosion, must not generate particles or vapor, and must not be adversely affected by regular wiping with water-based cleaners. Any interruptions in smooth surfaces that can collect particles should be eliminated, including holes, slots, and crevices. All interior and exterior corners should have larger than usual fillets and radii (Sematech standard 92051107A-STD). Polished exterior finishes are recommended wherever possible. Preferred materials and finishes include stainless steel, painted steel, and anodized aluminum. Elastomeric urethane, urethane enamels, and epoxy enamels are preferred for painted surfaces. All surfaces are to demonstrate 10^6 to 10^9 Ω/sq resistivity, per ASTM-D257 (see Section 3.2.3). Galvanized and decorative finishes are not recommended. A selection of suitable surface finishes and selection criteria are discussed below.

3.3.1 Common surface finishes

Anodized aluminum

During the electrochemical anodizing process the aluminum work piece is the anode, and the cathode is an inert electric conductor that does not react with the anodizing solution (carbon, lead, nickel, or stainless steel). Electrons are removed from the aluminum, allowing ions at the surface to react with water and form a hard, porous oxide layer up to 50 μm thick, half of which is added to the aluminum surface while the other half penetrates the surface (Alwitt, 2002). The process is largely controlled by the anodization bath, usually acid electrolytes (dilute sulfuric acid, oxalic acid, or phosphoric acid). The porous surface allows the aluminum to be dyed in a variety of colors. The oxide layer improves the scratch and wear resistance of aluminum and also provides corrosion resistance. Both properties reduce particle generation and make anodized aluminum a good surface finish, perhaps the most common finish for atmospheric cleanroom robots. The anodized layer does not peel or crack like organic coatings such as paint. Aluminum oxide is an electrical insulator and also has good thermal properties. However, for some applications its electrical insulation is a disadvantage for ESD control (Aalund and Mathia, 2007).

Powder coating

Wet paint is not recommended for cleanroom applications, because paint can wear and flake off, resulting in particle contamination. Powder coating, on the other hand, is suitable for cleanrooms. It provides a protective finish that is similar to wet paint in appearance, but is relatively hard and abrasion-resistant. The powder is a mixture of dry, finely ground particles of pigment and resin. It is applied to the surface, usually electrostatically, and is then cured with heat to form a uniform coat. Powder coating does not require a solvent. Thermoplastic powders and thermosetting powders are the common powder types. They can be applied to various materials, and generate a good coverage and durability compared to standard wet paint. In cleanroom robotics they are often used for aluminum and steel. A guideline for the coat thickness is at least 25 μm for non-aggressive atmospheric applications, and at least 60 μm for harsher environments. Note that standard powder coating is an electrical insulator and does not provide a dissipative surface. However, conductivity can be achieved with the addition of carbon or other conductive elements to the powder prior to curing.

Mechanically polished stainless steel

Mechanically polished stainless steel has a near mirror finish which is chemically inert, conductive, and excellent for controlling airborne molecular contamination (AMC). It has an attractive and clean appearance. In addition to its visual appeal, a polished stainless steel surface finish has other important properties:

- It is less susceptible than brushed stainless steel to accumulating particles that later could be released into the cleanroom.
- It is easy to clean and keep clean.
- It is only moderately susceptible to accumulating aggressive deposits that can initiate local corrosion.

Generally, a smoother surface provides better cleanliness, so polished stainless steel surfaces are common in cleanroom applications. In vacuum, uniformly smooth surfaces with a surface roughness of 0.2 to 0.4 μm are recommended (R. Bergner, personal communication, 2008). Disadvantages of polished surfaces include sensitivity to any cosmetic damage. The higher cost compared to brushed surfaces is also a limiting factor in the use of polished stainless steel.

Brushed stainless steel

Brushed stainless steel has a dull, chemically inert, and conductive surface finish that is good for controlling airborne molecular contamination (AMC). The cost of brushed stainless steel is competitive compared to polished stainless steel, and the dull surface finish is less sensitive to cosmetic damage. However, the increased surface roughness can provide cavities and voids in which particles can collect. It is recommended that a consistent brush line appearance be maintained at corners and panel intersections. Brushed finishes are used in cleanroom robotics, mainly in atmospheric and non-aggressive applications. Brushing is also used with aluminum when subsequent plating is applied, for example with nickel. A concern in ultra-clean environments can be the accumulation and later release of particles. Surface cleaning is more difficult with brushed than with polished surfaces. The ISO standard 1302:2002 specifies surface textures.

Electropolished stainless steel

Electropolishing is an electrochemical process that removes material from a metallic work piece. It may be viewed as the inverse of electroplating. A work piece is immersed in a liquid medium and subjected to a direct current which flows from the work piece, the anode, to the cathode, thus removing metal ions. The bath chemistry, temperature, time, and current density are controlled to produce a smooth, shiny coating, resulting in an attractive, crystalline metal surface that reduces particle contamination and adhesion. Unlike mechanical finishing systems, electropolishing does not have the potential to smear, bend, stress, or fracture a metal surface. However, the electrical process may leave a chemical residue, which must be removed to prevent AMC. Overall the surface has similar characteristics to that of mechanically polished stainless steel.

Cold rolled steel and urethane paint

Cold rolling is a common method of forming sheet metal. Cold rolled steel covered with liquid urethane paint is a low-cost alternative to stainless steel, but does not offer the same robustness or coverage as a powder coat. Cold rolling increases the hardness and strength of metal, with the side effect of an increased probability of cracks and fractures. The liquid paint usually prevents the release of particles from these cavities.

Nickel plating

Nickel plating is an adherent surface finish with some desired properties, such as electrical conductivity, chemical resistance, and resistance against abrasion, wear, and corrosion. During electroplating a nickel layer is deposited on the work piece. Nickel plating is common in atmospheric cleanroom robotics and offers an attractive surface finish with an appearance similar to chrome and stainless steel. Nickel-plated aluminum is a good alternative to anodized aluminum. It is suitable for harsh environments and for conductive surfaces: if ESD from charged substrates is a concern a nickel plated robot manipulator and base provide the needed grounding. Two plating methods are common: one uses electrical current for coating electrically conductive work pieces; the second is electroless nickel plating that can also be applied to non-conductive surfaces. Alternative plating materials are chrome, rhodium, and cadmium.

Everslick® and Parylene®

Non-metallic surface finishes include Parylene® and Everslick®. Both are ranked as good coatings for harsh environments. Surface treatments for harsh environments can be tested using a salt fog test per ASTM standard B117-97.

3.3.2 Selecting a surface finish

The choice of surface finish can influence the risk level of electrostatic discharge (ESD), electrostatic attraction (ESA), and airborne molecular contamination (AMC), and thus can influence product safety and yield. There are also economic factors related to cost, appearance, and brand recognition. Some industry standards influence the selection of surface finishes: the SEMI standard E10 recommends surface properties for manufacturing equipment. This section reviews available and emerging surface finishes for typical cleanroom applications (Aalund and Mathia, 2007):

- Mechanically polished stainless steel
- Electrically polished stainless steel

- Brushed stainless steel
- Powder coat over steel and aluminum
- Cold-rolled steel with liquid urethane paint.

Table 3.7. Relative importance of selection criteria for surface finishes.

Rank	Evaluation and selection criterion	Relative importance
1	Electrostatic discharge and attraction (ESD, ESA)	100%
2	Airborne molecular contamination (AMC)	81%
3	Cost of assembly	77%
4	Cost of materials	61%
5	Appearance and brand recognition	61%

The evaluation of these finishes using suitable selection criteria showed that selecting the best finish for a given application is not always straightforward. The decision must balance several factors, including cost, compliance with regulatory requirements, productivity, quality, and customer requirements. In low volumes the material costs are similar for all finishes considered. For high volumes, however, powder coats tend to have a cost advantage. On the other hand, polished stainless steel is a superior material with respect to AMC, ESD, and ESA, although dissipative powder coats that comply with the SEMI standard E78 can neutralize the ESD and ESA advantages. Powder coats also offer various colors that can influence tool appearance. In short, the technical advantages of some finishes are offset by the economic advantages of others, and the best choice may be application-dependent.

A set of selection criteria was established based on ITRS guidelines, relevant and emerging standards in electronics manufacturing, and feedback from tool manufacturers. The criteria were ranked by manufacturers of equipment front end modules (EFEMs). EFEMs are mini-environments used in the semiconductor front-end industry for processing 300 mm wafers. They were selected for this study because they provide a controlled sample set, are governed by SEMI standards, and are widely used for automated wafer transport between wafer cassettes and manufacturing tools. The selection criteria for surface finishes are ranked by their relative importance (the average number of votes). The highest ranking was arbitrarily set to 100%. Table 3.7 shows the ranking. The table suggests that product safety is the biggest concern: ESD, ESA, and AMC are ranked 1 and 2. Cost is the next criterion: assembly and material costs are ranked 3 and 4. Appearance and brand recognition is ranked 5.

The selection criteria are described below.

Selection criteria

Selection criterion 1: electrostatic discharge and attraction

Electrostatic charge in semiconductor manufacturing tools poses a high risk to product safety and therefore to wafer yield and profitability. This is caused by two primary effects: electrostatic discharge (ESD) and electrostatic attraction (ESA). See Section 3.2.

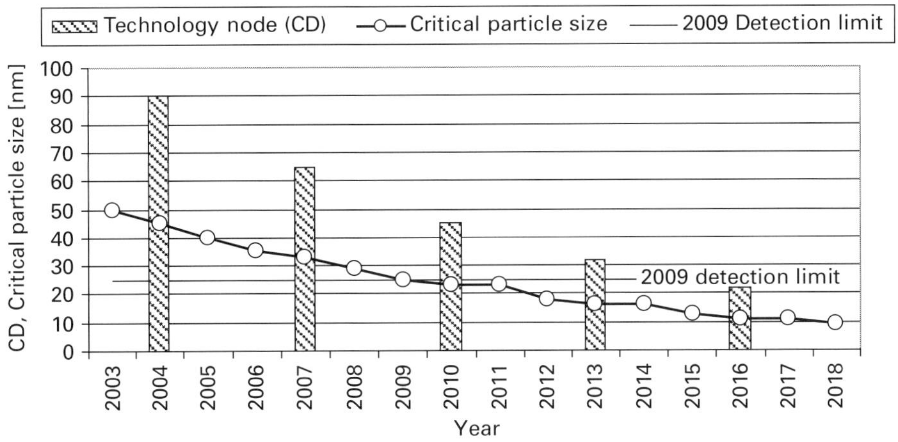

Figure 3.3 Projected technology nodes and critical particle sizes (Aalund and Mathia, 2007).

Selection criterion 2: atmospheric molecular contamination

With the feature sizes of modern electronics shrinking below 0.1 μm, prevention of airborne molecular contamination (AMC) in manufacturing environments is critical. The ITRS is promoting a schedule of increasingly stringent cleanliness milestones through 2018, including a continuously decreasing critical particle size and wafer surface contamination. The critical particle size is about 50% of the feature size of the semiconductor devices manufactured on a wafer. Figure 3.3 illustrates the ITRS roadmap for the technology nodes (critical dimensions, CD), together with the associated critical particle sizes. For example, in 2010 the CD is 45 nm and the critical particle size is 23 nm. As of 2009 the detection sensitivity of commercially available metrology tools is limited to defect sizes of about 25 nm (example: KLA-Tencor's Surfscan SP2). The specified cleanliness (AMC limit) is 100 m^{-3} for particles with the critical particle size or larger.

Selection criteria 3 and 4: cost of assembly and materials

Cost is a major factor in the semiconductor business equation. How do surface finishes and quantity influence the cost of typical EFEM parts and components? It has been shown that mirror-polished stainless steel, anodized aluminum, and aluminum coated with urethane paint are among the most expensive finishes. The cost of stainless steel with urethane is slightly lower. However, dissipative powder coats over aluminum are an attractive alternative.

Selection criterion 5: appearance

The appearance of a tool can often influence the selection of a surface finish for visible parts of a cleanroom tool or robotic system. Mirror finishes or specific colors are regularly specified for external surfaces.

Data analysis and ranking

The above case study established the ranking of the selection criteria for surface finishes listed in Table 3.8. Mechanically polished stainless steel was ranked as the best finish

Table 3.8. Overall ranking of surface finishes.

Rank	Surface finish
1	Mechanically polished stainless steel
2	Powder coat on aluminum
3	Electrically polished stainless steel
4	Brushed stainless steel
5	Urethane paint on cold-rolled steel

with respect to ESD, ESA, and AMC. Powder coats were ranked best with respect to cost. Stainless steel has the advantage of being conductive, but dissipative powder coats and urethane paint that comply with SEMI standard E78 neutralize that advantage. In summary, the selection of a surface finish for cleanroom robots is not always straightforward and should be part of the product development process.

3.4 Clean drive trains

A cleanroom-compatible drive train does not degrade air cleanliness by generating particles beyond the cleanroom specification. The vast majority of industrial robots utilize electromechanical drive trains with electric motors, so the 'primary energy' referred to in Definition 3.2 is electrical energy.

Definition 3.2: A robot drive train is the set of moving parts and components that transforms primary energy into kinetic energy and delivers it to the robot's interface with the outside world to perform a commanded manipulation.

In substrate handling the robot interface with the outside world is the end-effector. The drive trains considered here include electric motors, transmissions, belts and pulleys, linear and rotary bearings, and mechanical parts. Power amplifiers and position sensors (encoders) are considered part of the control system.

3.4.1 Typical design concept

Figure 3.4 shows a cross section of a typical substrate-handling atmospheric robot. The robot has a SCARA-type arm (see also Figure 2.7). Most substrate-handling robots in electronics manufacturing, in particular in the semiconductor industry, employ this or similar concepts. Figure 3.4 illustrates the main components of the drive train:

- One motor per axis of motion
- *Z*-axis assembly with vertical prismatic joint
- SCARA-type arm with two arm links and three revolute joints, and three belts with pulleys.

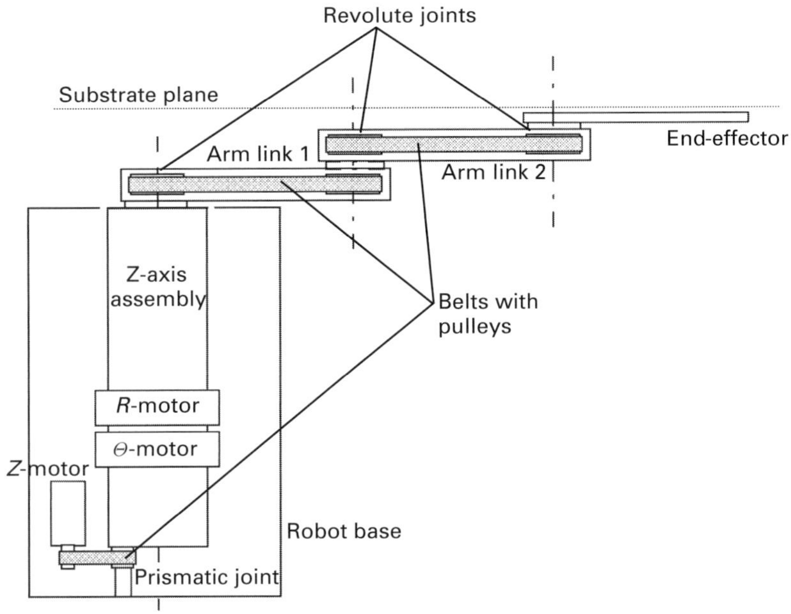

Figure 3.4 Cross section of a SCARA-type robot and drive train concept.

The axis of motion for radial arm extension and retraction is the R-axis. The R-motor driving the R-axis has an opening for the drive shaft that transfers the motor torque to the first arm link, and to the second arm link with belts and pulleys. The kinematics is discussed in Section 5.4.4. As a special case, the robot arm in Figure 3.4 becomes a cylindrical robot if its two-link arm is constrained to radial motion only. See Example 5.6. The Z-motor elevates the entire Z-axis assembly, including the drive train for the R-axis, using a prismatic joint (typically a lead screw). The transmission from Z-motor to the prismatic joint is implemented with a belt and two pulleys. The Z-axis assembly also comprises the θ-motor for arm rotation. The risk of particle contamination resulting from the moving parts of drive trains is evident. However, it has been demonstrated that good engineering design practices can achieve ISO Class 1 cleanliness (Pentagon, 2002; Pentagon, 2004). The following is a list of design guidelines that support cleanroom-compatible robot designs.

3.4.2 Guidelines for clean drive trains

The following are basic guidelines for the design of clean drive trains. Several standards also address clean designs, for example the Sematech standard 92051107A-STD.

Maintainability

A drive train should be easy to access and to maintain, including for cleaning and repairs, so that an acceptable contamination level can be quickly restored if necessary.

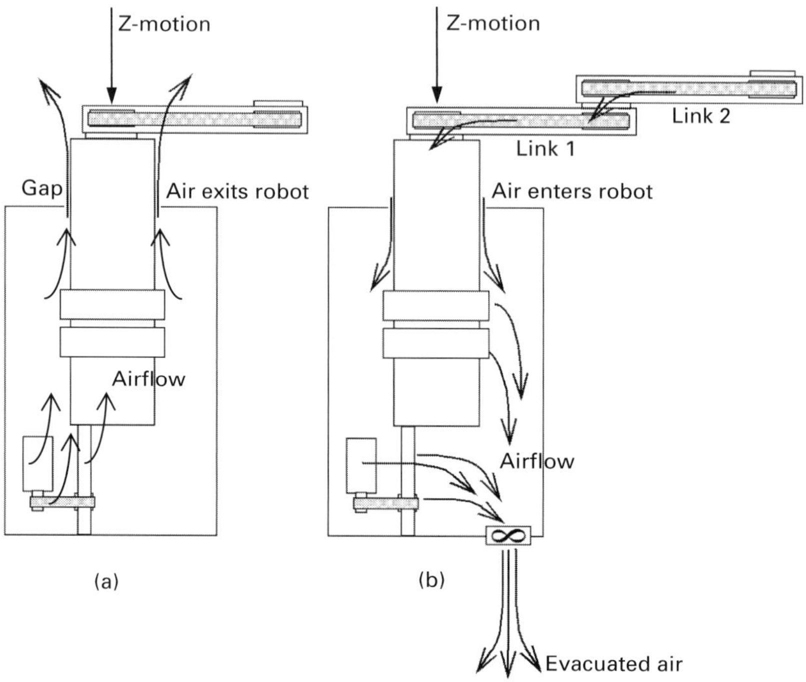

Figure 3.5 (a) Generation of airborne particles; (b) evacuating particles from the robot base.

All parts below substrate

All parts, most importantly moving parts, must be located below the substrate plane at all times (Figure 3.4). This prevents released particles from falling onto the substrate.

Evacuating generated particles

Particles generated inside the robot should be evacuated through a dedicated exit near the bottom of the robot base. Figure 3.5a illustrates how particles are 'pumped out' of the robot base when the *Z*-axis assembly retracts into the base. Without controlled evacuation the air displaced by the *Z*-axis assembly exits the robot base through any available openings, particularly through the gap between the *Z*-axis assembly and the top of the robot base. This potential contamination occurs in the vicinity of the substrate and poses a substantial risk to product yield. The 'pumping' effect can be avoided with an air pressure inside the robot base that is slightly below that of the ambient atmosphere. The necessary airflow can be provided by a small fan, as is shown in Figure 3.5b. The contaminated air exits the robot base through the bottom plate, in the direction of the tool's laminar airflow. This exit location not only maximizes the distance between the evacuated particles and the product, but also avoids disruption of the laminar, top-to-bottom airflow used in some mini-environments. Laminar airflow is intended to transport airborne particles away from the substrates, but vertices could elevate particles back up to the product. The air volume flow rate provided by the fan must exceed the maximum possible air volume flow created by the *Z*-axis assembly at maximum negative *Z*-velocity. If that is the case a continuous, small flow of clean air enters the robot base through the gap between the robot base and the *Z*-axis assembly, instead of contaminating the substrate.

Moving parts

Moving parts generate particles. Minimizing the number of moving parts is an important design objective for cleanroom robots. Before the materials for moving parts are selected the abrasion and wear characteristics, as well as the thermal and mechanical limits, of candidate materials must be reviewed and compared against the specified operating conditions. High-quality, wear-resistant coatings can reduce friction and particle generation. For example, the 'PV value,' the product of unit pressure and surface velocity, can be a limiting material parameter. Titanium nitride and titanium carbide coatings on steel reportedly have produced good results. Synergistic coatings, that is, metal infused with a low-friction polymer, also can give good results. Of course, the cost of these materials must be justified. Moving parts in the proximity of the substrate should be enclosed in order to prevent particles from entering the robot's clean environment. Seals for bearings are a good example (see below).

Motor selection and direct drives

Brushed motors offer inexpensive and proven technology, while brushless servo motors can generate fewer particles. If the tight space requirements allow, direct-drive mechanisms may be considered. Drive trains with direct-drive joints avoid contaminating belts and pulleys and are good choices for cleanroom robots, if the results justify the extra cost. (Note: the term 'direct drive' is not always clearly defined. It is sometimes used even with belts present in the robot arm.)

Belts and pulleys

Belts can be a significant source of particle contamination, which can be reduced if suitable belt and pulley materials are used, and a belt and pulley combination is specifically selected based on the expected operating conditions, including forces, speed, duty cycle, temperature, and other parameters. Belt manufacturers offer recommendations for belt materials and size, belt/pulley alignments, and belt tension.

Gearheads

Gearheads offer an alternative to torque transmission via belts and pulleys. The trend towards automation motivated the use of gearheads even in cleanrooms and moderate vacuum, despite the fact that cleanrooms are considered an 'extreme' environment for gearheads. Cleanroom-compatible and vacuum-compatible gearheads are made with stainless steel, special seals, and suitable lubricants. The materials used must withstand potentially aggressive electronics manufacturing processes and cleaning solutions.

Lubricants

Lubricants for moving parts such as bearings, lead screws, and gearheads must be nonvolatile and their use must follow strict procedures. Dry powder lubricants and silicone lubricants should be avoided.

Smooth motion trajectories

Smooth motion trajectories reduce friction and wear and as such particle contamination. Rapid motion, heavily loaded motion, and contact or impact with the robot's environment should be avoided.

Vibration control

Vibration can generate particles "simply by shaking them out of tooling components" (Sematech), and must be avoided. Examples are particles on process chamber walls, or particles generated by substrates that slide into cassettes. More severe effects include the disintegration of composite materials from vibration-induced stress. The Sematech standard 92051107A-STD recommends, particularly for vacuum robots, that vibration levels in the product vicinity be barely noticeable. The suggested velocity limit v_{max} and acceleration limit a_{max} are, depending on the vibration frequency f:

$$\begin{aligned} f &< 100\ \text{Hz}: v_{\max} = 7.6 \cdot 10^{-4}\text{m} \cdot \text{s}^{-1} \\ f &> 100\ \text{Hz}: a_{\max} = 0.49\text{m} \cdot \text{s}^{-2} = 0.05\text{g}. \end{aligned} \tag{3.3}$$

Substrate-handling robots must not contribute vibration to the process chamber, nor to any portion of the substrate load/unload system.

Internal cabling

All wires and cables should be internal to the robot and arm. Internal cabling allows for the controlled evacuation of airborne particles that are generated by the inevitable rubbing and flexing, and the resulting abrasion and wear, as well as the possible outgassing of insulation materials. Wires and cables should be fastened in fixed locations if possible. Cleanroom-compatible insulation materials are commercially available. External cables, if required, are usually fixed in tool locations that are easy to evacuate; an example is cables to and from the robot controller. However, if linear horizontal displacements of the entire robot are needed, this additional robot axis can be implemented with a 'linear track.' Typical motion ranges are 0.5 to 2.0 m, which requires external cabling and precautions to minimize particle generation, for example with cleanroom-compatible cable carriers (Pentagon, 2001).

Seals for revolute joints

Bearings are the primary components of revolute joints and can be isolated with sealing devices in order to prevent contamination in both directions: lubricant and particles could exit the bearing, or foreign matters could enter the bearing. Here the focus is on preventing particle contamination of the robot's environment. Contact and non-contact seals are available for this purpose (Figure 3.6):

- Contact type seals press a sealing lip against the rotating shaft, effectively excluding contaminants and retaining the bearing lubricant (Figure 3.6a). However, the contact with the shaft increases friction and can reduce the maximum speed of an axis of motion.

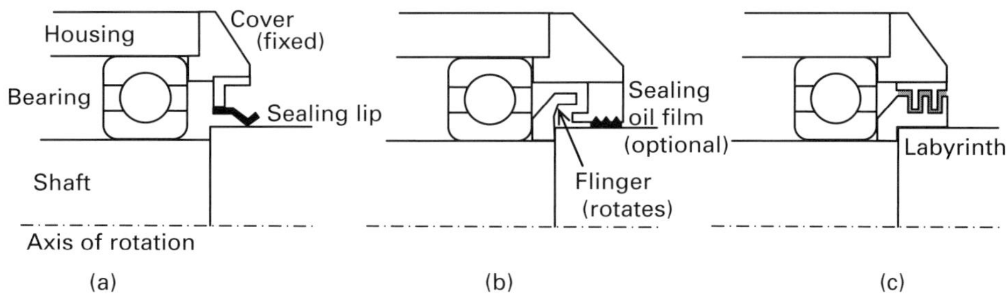

Figure 3.6 Bearing seals: (a) contact type, (b) non-contact flinger, (c) non-contact 'labyrinth'. Only one half of the bearing assembly and shaft is shown.

- Non-contact type seals include flinger, labyrinth, and oil-groove types. They are easy to mount and dismount and generate minimal friction. The flinger uses centrifugal force and a small air stream to retain lubricant and particles (Figure 3.6b). The radial labyrinth type is well suited for high speeds (Figure 3.6c).

3.5 Arm compliance

Robot arm compliance (elasticity, the inverse of arm stiffness) in the semiconductor, FPD, hard disk, and solar manufacturing industries is primarily required for handling flat substrates in horizontal planes with a specified vertical tolerance. In wafer processing, for example, that tolerance is ±0.5 mm for vacuum cluster tools (SEMI standard E21-94). The overall robot stiffness, the combined stiffness of robot base and arm, must maintain this requirement under various operating conditions and wafer heights. The wafer height is defined as the distance from a reference plane to the wafer plane with respect to a wafer's bottom surface. The reference plane is usually the lowest possible wafer plane, either the robot's lowest vertical position ($Z=0$) or the bottom position of a wafer cassette. For SCARA-type robots compliance is determined by the combined structural design of the vertical Z-column and the horizontal arm. A challenge with the arm design can be the space limitations imposed by narrow tool doors and load locks. Light weight is also desired to support fast robot dynamics and high throughput with moderate motor torques and power consumption. This section first defines the forces, moments, and arm parameters, then presents a conceptual design for a revolute joint and a SCARA-type arm for 300 mm wafer handling.

3.5.1 Forces and moments

The reaction forces and moments of the designed SCARA-type arm illustrated in Figure 3.7 are analyzed here. The design objectives are sufficient arm stiffness for a single 300 mm wafer and an edge-gripping end-effector, as well as the elimination of reaction moments. First the reaction forces and moments acting on the three revolute joints are formalized and analyzed; this is necessary in order to specify the mechanical

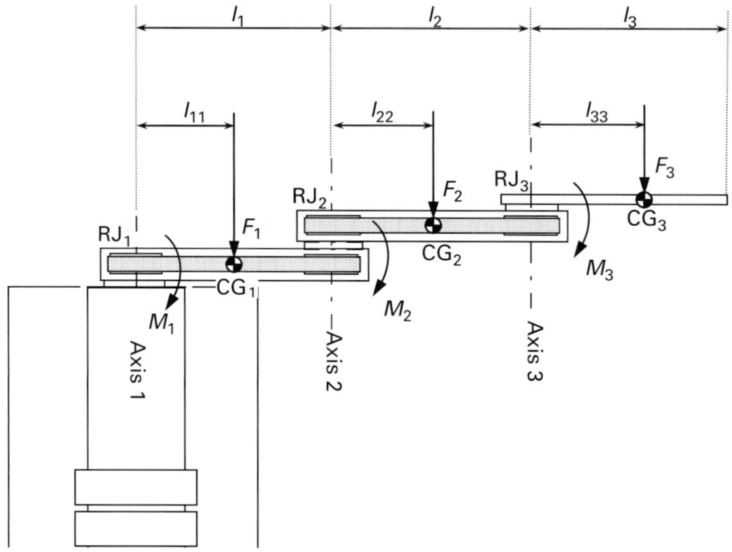

Figure 3.7 Forces and moments acting on a SCARA-type robot arm.

parameters of the revolute joints and arm links. Vertical motion, provided by the independent Z-axis of motion, does not affect the arm design. The three revolute joints, the 'shoulder,' 'elbow,' and 'wrist' in Figure 3.7, are labeled RJ_1, RJ_2, RJ_3, respectively. The two arm links and the end-effector have effective lengths l_1, l_2, l_3, defined as the distance between the axes of rotation of the respective revolute joints. The centers of gravity of the two links and the end-effector (plus wafer) are labeled CG_1, CG_2, CG_3. The resulting gravitational forces are F_1, F_2, F_3. The distance between the i-th joint axis and the i-th CG is l_{ii}.

Designing the three revolute joints, including the selection of bearings, requires the values of the reaction forces R_{ii} generated by the moments M_1, M_2, M_3. The moments in turn are generated by the gravitational forces acting on the i-th arm link, $M_i = F_i \cdot l_{ii}$. The largest moment occurs at the shoulder joint RJ_1 and requires an appropriate joint RJ_1 and inner link 1. The inner link of commercial wafer-handling robots is up to 50 mm thick. The vertical dimensions of the outer link and the end-effector can be critical design parameters for reaching through the narrow openings of process tools and load ports.

3.5.2 Design concept for revolute joints

The conceptual design of a revolute joint that eliminates reaction moments is presented here.

Ball bearings

Bearings support a shaft or housing and permit free motion about the bearing's shaft. Radial loads are perpendicular to the shaft. Axial (thrust) loads are directed parallel to the shaft. Any deviation from these two directions results in a load moment. Various bearing types are available for handling radial, axial, and moment loads. Ball bearings are the most common type and support both radial and thrust loads. They are typical for applications with constant, light to medium loads and are the main components of revolute joints (I. Kremerman, personal communication, 2009). Revolute joints provide the basis for kinematic chains of rotary linkages. Such structures allow for smaller sweep radii and smaller footprint requirements than kinematic structures based on prismatic joints. These are desired features of handling robots that operate in tight spaces.

The advantages of ball bearings for cleanroom robotics include:

- Low particle generation due to low dynamic friction
- Ease of sealing for containing generated particles
- Small size (low profile) in the axial direction
- Low static friction
- Moderate lubrication requirements
- High accuracy of rotation (low values of run-out)
- High operating temperatures
- Suitability for vacuum applications.

Potential disadvantages of ball bearings for cleanroom robotics include:

- Relatively high cost
- Sensitivity to installation errors
- Relatively large radial dimensions.

Sealing techniques for bearings are illustrated in Figure 3.6.

Design objectives

The design objectives for the revolute joints used for the robot arm in Figure 3.7 include:

- Maximize the distance between bearings to minimize reaction forces.
- Use preloaded angular contact bearings or deep-groove bearings to minimize backlash and angular joint deflection.
- Avoid cantilevered bearing installations to minimize angular deflection.

The design concept presented in Figure 3.8 meets the objectives and eliminates reaction moments. Only the reaction forces R_{ii} within the bearings apply. Figure 3.8 shows a free-body diagram of the i-th revolute joint of the arm in Figure 3.7. It utilizes two bearings: B_{i1} is a radial bearing that only generates a radial (here: horizontal) reaction force R_{i1x}, while bearing B_{i2} only generates the axial (here: vertical) reaction force R_{i2z}. The vertical and horizontal rollers in the figure illustrate this concept. Among the many variations of revolute joints, this rigid design is particularly suited for high loads. Several implementations are possible: bearing B_{i2} can be realized with two angular contact bearings (in a preloaded face-to-face configuration), one four-point contact ball bearing, or one roller bearing. This two-point support eliminates undesired reaction moments and emphasizes only reaction forces:

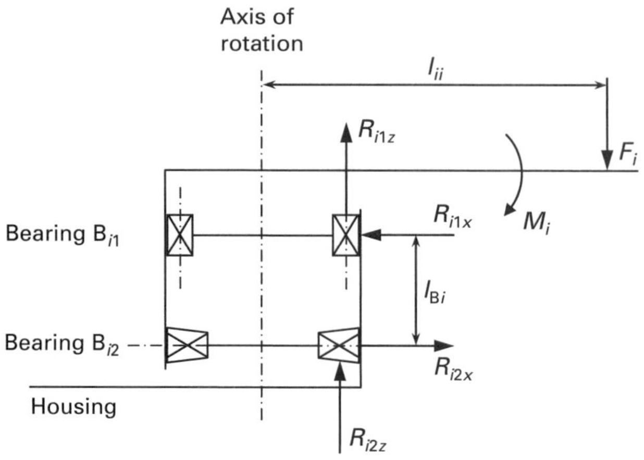

Figure 3.8 Free-body diagram of the i-th revolute joint RJ_i.

- It is less sensitive to machining and assembly inaccuracies than other designs.
- The radial bearing B_{i1} provides horizontal rigidity. The reaction force R_{i1x} prevents radial backlash.
- The preloaded, angular-contact bearing B_{i2} provides vertical rigidity. It eliminates axial backlash by directing the reaction force R_{i2z} against the vertical load F_{iz}. It also provides the horizontal reaction force R_{i2x} while allowing free rotation.

The notation for the i-th revolute joint in Figure 3.8:

l_{Bi}	… vertical distance between bearings B_{i1} and B_{i2}
R_{i1x}, R_{i2x}	… horizontal reaction forces at bearings 1 and 2
R_{i1z}, R_{i2z}	… vertical reaction forces at bearings 1 and 2

3.5.3 Force and moment analysis

The following analysis quantifies the reaction forces per revolute joint, using well-known engineering methods. The reaction moments are eliminated by the above joint design. The results will provide the basis for the joint specification and bearing selection. The analysis is based on the free-body diagram in Figure 3.8. Note that $R_{i1z} = 0$ because the radial bearing B_{i1} does not provide an axial reaction force. Similarly, R_{i2z} depends on the mechanical configuration of the preloaded angular contact bearings. This follows from the sum of all axial and horizontal forces. The free-body diagram in Figure 3.9 utilizes this joint concept for all three revolute joints, as is shown in the arm design in Figure 3.10. All forces, moments, and reaction forces are shown.

The sum of all moments acting on each joint is zero, therefore

$$\begin{aligned} M_1 &= F_1 \cdot l_{11} + F_2 \cdot (l_1 + l_{22}) + F_3 \cdot (l_1 + l_2 + l_{33}) \\ M_2 &= F_2 \cdot l_{22} + F_3 \cdot (l_2 + l_{33}) \\ M_3 &= F_3 \cdot l_{33}. \end{aligned} \tag{3.4}$$

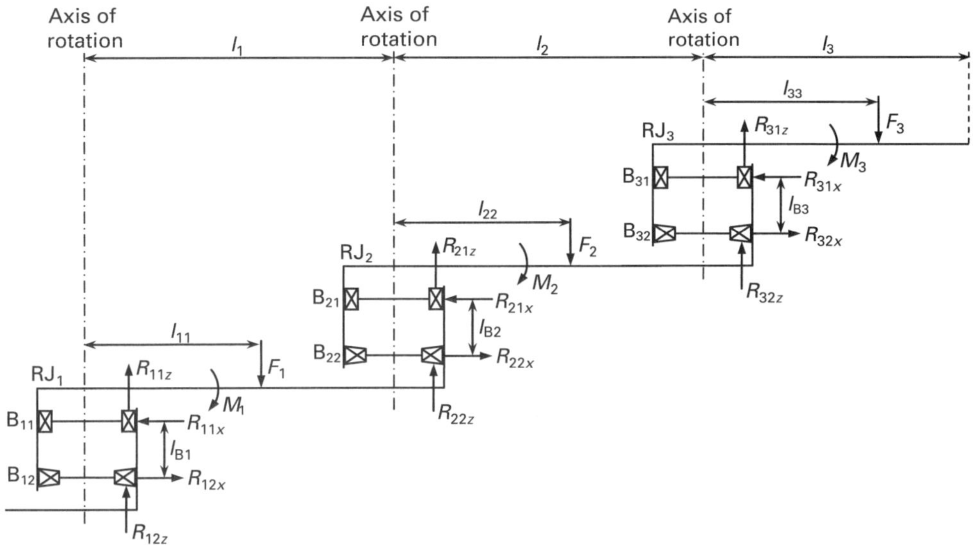

Figure 3.9 Free-body diagram of a SCARA-type arm design.

Similarly, the sums of all axial (vertical) forces are zero:

$$\begin{aligned}
&R_{11z} + R_{12z} - F_1 - F_2 - F_3 = 0 \\
&R_{21z} + R_{22z} - F_2 - F_3 = 0 \\
&R_{31z} + R_{32z} - F_3 = 0.
\end{aligned} \tag{3.5}$$

The sum of the total forces and moments at bearings B_{i1} and B_{i2} of the i-th revolute joint RJ_i ($i = 1,2,3$) are zero:

$$\begin{aligned}
&R_{i1z} = 0 \quad \text{(by design)} \\
&R_{i1x} \cdot l_{Bi} + M_i = R_{i2x} \cdot l_{Bi} + M_i = 0, \text{therefore} \\
&R_{i1x} = R_{i2x} = -\frac{M_i}{l_{Bi}}.
\end{aligned} \tag{3.6}$$

See Example 3.1 for a numerical evaluation of the above revolute joint concept. The calculated reaction forces allow the selection of suitable bearings for the revolute joints. All common ball bearing types are used in cleanroom robotics: angular contact bearings, single deep-grove bearings, cross-roller bearings, single and double row/radial contact ball bearings, double row/spherical ball bearings, single and double row angular contact ball bearings, double row/angular contact pre-loaded ball bearings, and four-point contact radial ball bearings.

Once the mechanical arm design is completed, cleanliness and ESD/ESA requirements can be addressed. Refer to Sections 3.1 and 3.2 for cleanliness guidelines and ESD prevention guidelines. The overall cost of arm materials and assembly labor should be regularly reviewed during the design process.

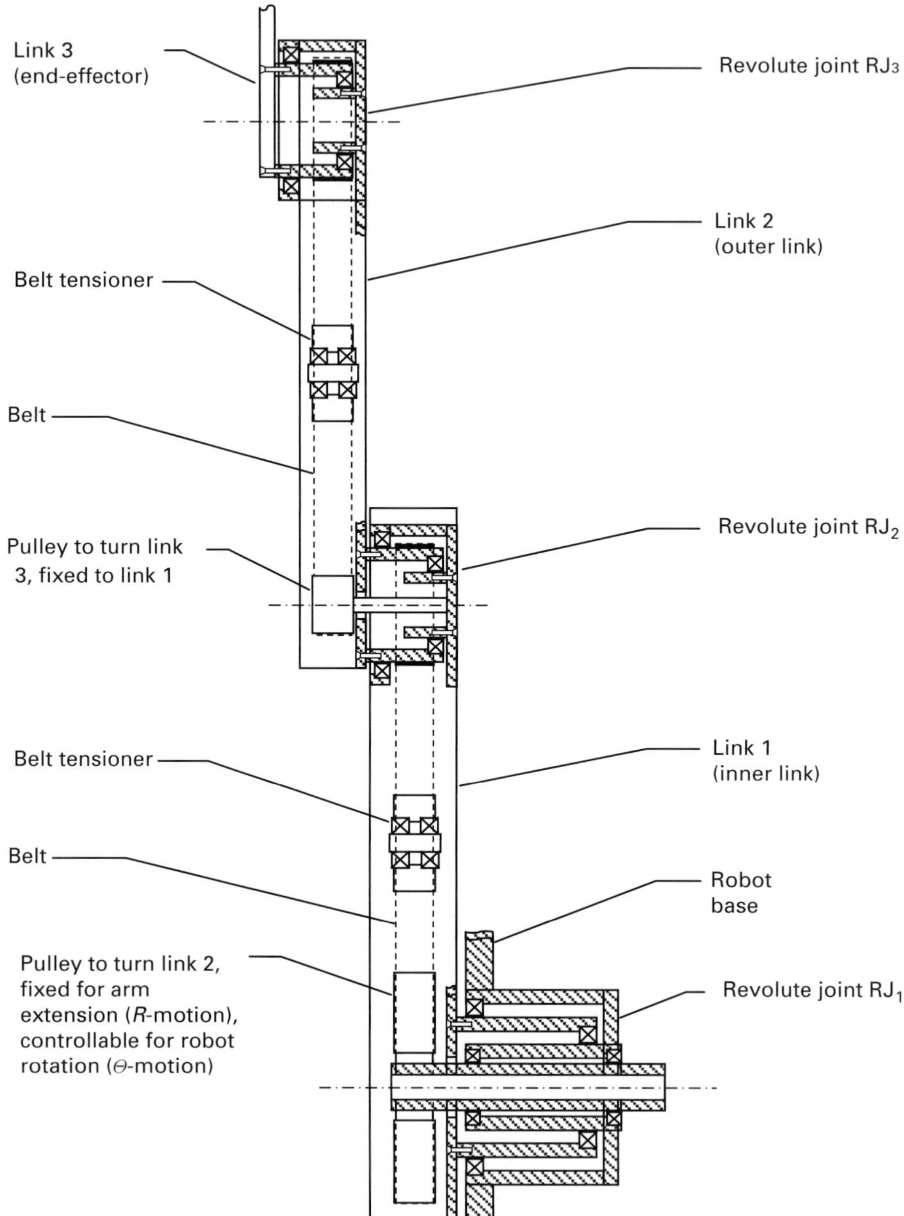

Figure 3.10 Cross section of the arm design in Example 3.1.

Example 3.1: reaction forces and mechanical design of a robot arm

This example presents the reaction force analysis of a SCARA-type arm for 300 mm wafer handling:

- The robot moves the substrate with a vertical tolerance of ±0.5 mm in the horizontal wafer plane. Refer to the SEMI standards E21-94 and E21.1-1296.
- The vertical dimension of end-effector and wafer combined is small enough to enter process tools and wafer cassettes. For example, the wafer slots in a front-opening unified pod (FOUP) for 300 mm wafers measure 10 mm.
- The weight of the robot arm should be sufficiently small to allow the fast manipulator dynamics needed for high wafer throughput.

The forces and moments are analyzed using Equations (3.4) to (3.6). Table 3.9 specifies the arm parameters shown in Figure 3.9. The load force $F_i = m_i \cdot g$ depends on the mass m_i of the i-th link and the gravitational acceleration $g = 9.81\ \mathrm{m} \cdot \mathrm{s}^{-2}$. The end-effector load F_3 includes the weight of a 300 mm wafer, about 2.5 N. The resulting load moments and reaction forces per joint are listed in Table 3.10.

Figure 3.10 shows the cross section of the designed three-link SCARA-type robot arm. The end-effector (link 3) is partially shown. The figure emphasizes the revolute joint concept in Figure 3.8, which is realized with two ball bearings per joint: one common radial bearing for horizontal stiffness (top bearing B_{i1}), and one preloaded four-point contact bearing for both horizontal and vertical stiffness (bottom bearing B_{i2}). The same joint design is utilized for the shoulder, elbow, and wrist joints. The inner link is designed to be stronger than the outer link. The larger reaction forces acting on the shoulder joint RJ_1 are reduced with an increased length $l_{\mathrm{B}1}$ between the top and bottom bearings B_{11} and B_{12}. An edge-gripping end-effector, heavier than blade-type end-effectors and discussed in Section 3.6, is assumed. This arm design is one of several possible solutions that provide the desired compliance.

The pulleys and belts can be used as a mechanical constraint that connects both arm links such that radial motion along the R-axis requires only one motor. This is accomplished with specific gear ratios between the inner and outer link. See Example 5.4 for details. □

Table 3.9. Arm design in Example 3.1: parameters for the i-th arm link.

i	Length l_i (mm)	Length l_{ii} (mm)	Length $l_{\mathrm{B}i}$ (mm)	Mass m_i (kg)	Force F_i (N)
1	178	80	40	0.765	7.5
2	178	90	13	0.401	4.0
3	295	145	7	0.734	7.2

Table 3.10. Reaction forces in Example 3.1.

	Revolute joint		
Parameter	RJ_1 ($i=1$)	RJ_2 ($i=2$)	RJ_3 ($i=3$)
M_i (N·m)	5.28	2.69	1.04
R_{i1z} (N)	0.00	0.00	0.00
R_{i2z} (N)	18.70	11.20	7.20
R_{i1x} (N)	132.00	206.58	149.14
R_{i2x} (N)	132.00	206.58	149.14

3.6 End-effectors

End-effectors provide the interface between a robot and the outside world. The term 'end-effector' refers to a tool at the end of the robot manipulator. While industrial robots are flexible devices designed for a range of applications, end-effectors are dedicated for specific robot tasks. They may be equipped with sensors to aid the robot in locating or positioning work pieces. Examples of end-effectors are welding guns, spray guns, grinders, and grippers. In electronics manufacturing various end-effectors are used for handling substrates, photo masks, flat panel displays, hard disks, and solar panels. Two end-effector types, edge-grippers and vacuum blades, are discussed here. Typical grippers utilize electromechanical or pneumatic actuators. However, many end-effectors for wafer handling in atmospheric applications simply use vacuum suction. Here the design objectives include:

- Operator safety: laser scanners are sometimes embedded in end-effectors for sensing and locating substrates. Laser safety measures avoid eye injuries and are provided by Class 1 lasers per IEC standard 60825-1.
- Cleanliness: the end-effector should not contaminate the substrate with particles. This applies to surface contamination through contact with the substrate, and through airborne particles generated by gripper mechanisms.
- Throughput: the end-effector should support a specified throughput by enabling fast handling without compromising the robot's substrate positioning repeatability.
- Flexibility: the mechanical and electrical interface between robot and end-effector should allow the use of grippers and vacuum end-effectors.

Figure 3.11 shows an end-effector for 300 mm wafers that uses vacuum suction to hold the wafer. This is called a 'paddle-type' or 'blade-type' end-effector and is typically made of anodized aluminum. Small PEEK pads make contact with the wafer backside. The top view in the figure shows the two concentric rings of the vacuum chuck and the mechanical interface with the robot arm, a circular bolt pattern. The design includes a 'brake-the-beam' laser sensor for wafer scanning. Note the amplifier compartment in the cross section A-A, and the machined channels for the fiber optics. The line between the two paddle tips indicates the laser beam. A typical design of an edge-gripping end-effector for 300 mm wafers is shown in Figure 3.12.

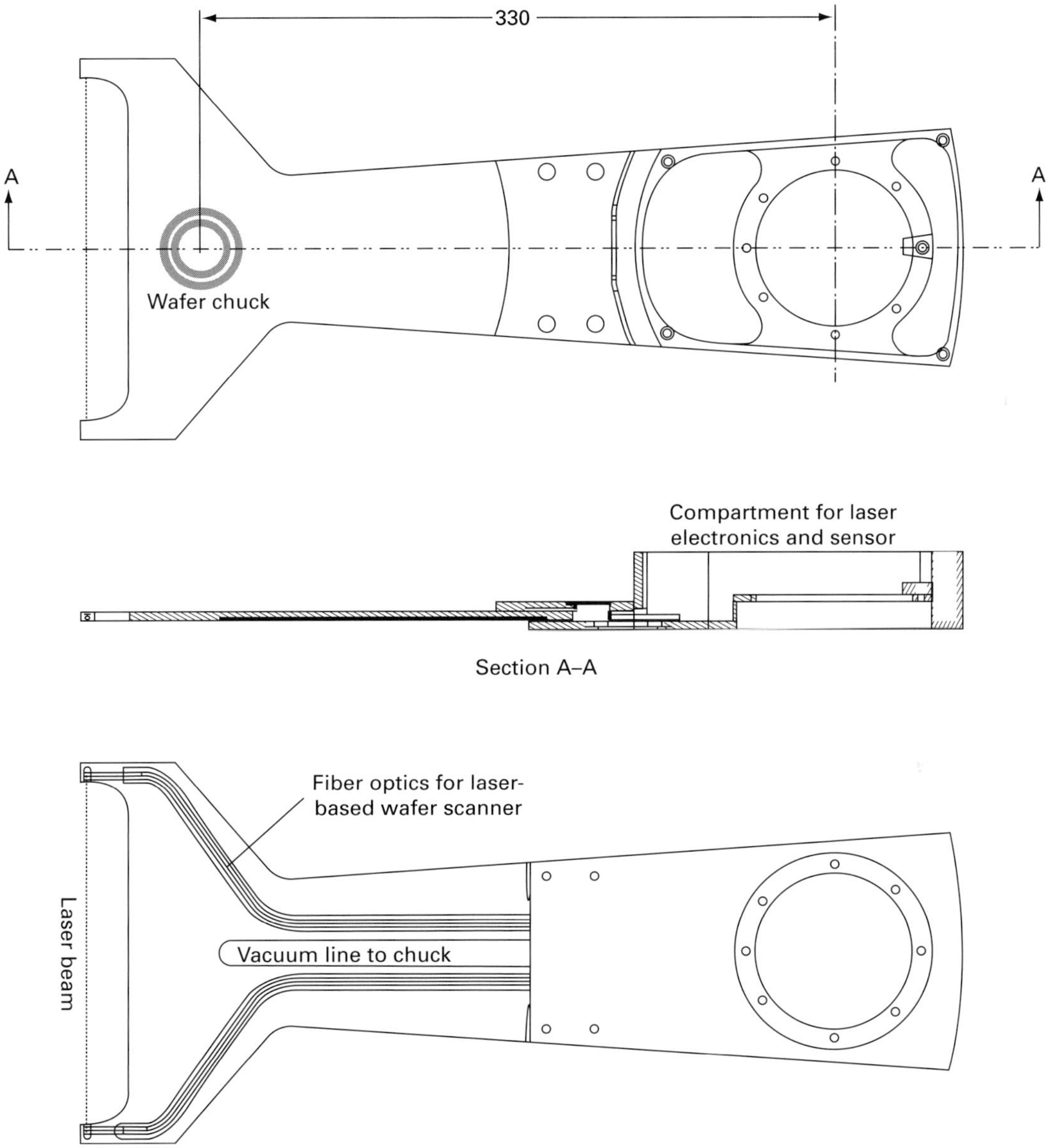

Figure 3.11 End-effector with vacuum suction for 300 mm wafer handling. Dimensions are in mm.

A robot detects wafers in a cassette during vertical motion. A brief interruption of the laser beam indicates a wafer at that position. This process is called 'wafer scanning' or 'wafer mapping.' Reflective laser scanners are also used: the laser beam is reflected by the wafer edge and then sensed. However, this method can be sensitive to the reflectivity of the wafer material and the shape of the wafer edge. The wafer cannot be detected if the reflected energy reaching the sensor is insufficient. A laser scanner should be able to detect 'cross slots,' that is, a substrate that is not stored horizontally but in two adjacent

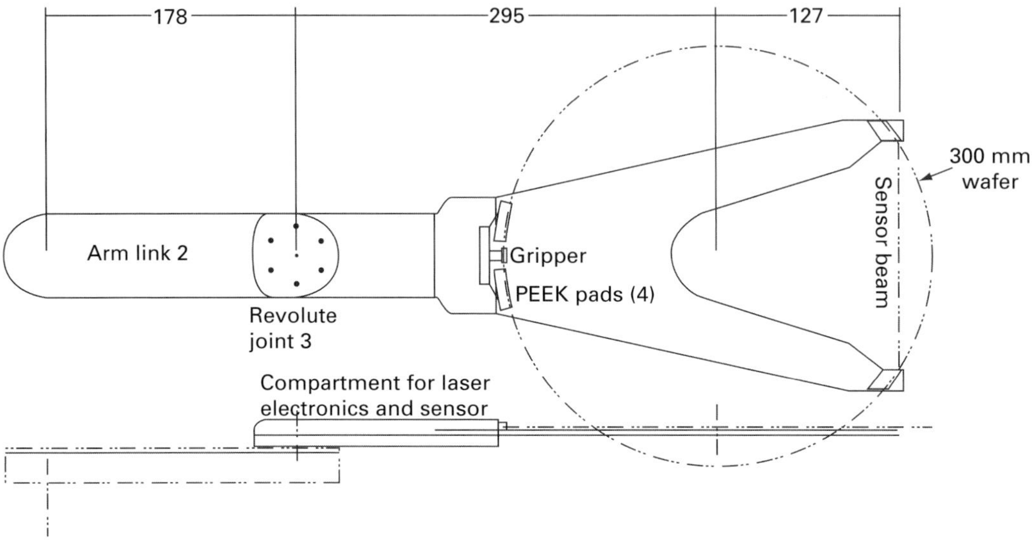

Figure 3.12 Edge-gripper design for 300 mm wafer handling. Dimensions are in mm.

slots in a storage cassette. Another challenge can be to identify two wafers in one slot. A successful wafer pickup should be verified with a wafer presence sensor.

Wafer sensing can be accomplished with a vacuum sensor in the vacuum line inside the robot base. Only clean, filtered air should be used for vacuum suction, otherwise the back-splash of air when opening the valve can contaminate the wafer backside. For edge-gripping end-effectors a gripper position feedback can be used to indicate the presence of a wafer. A flexible robot design allows for various end-effector types and provides appropriate vacuum and wiring for sensor and control signals, as well as power for gripper actuators.

3.7 Robot assembly and handling

It is difficult to remove particles, ionics, and organics once they have accumulated on the surfaces of robotic systems. It is therefore important to minimize contamination beginning with the robot assembly. The semiconductor industry provides guidelines for clean assembly areas, for example the Sematech standard 92051107A-STD. The following fundamental practices are recommended:

- Assembly, test, and preparation for shipment should take place in ISO Class 5 cleanliness or better.
- Procedures for minimizing contamination should be regularly promoted and trained.
- All parts and components must be cleaned and ready for assembly before entry to a clean assembly area. Machining of any kind is prohibited in cleanrooms.
- All tools and equipment must be cleaned prior to entry to the cleanroom, and must be cleaned regularly afterwards.

- Cleaning includes vacuum cleaning, blowing with clean, filtered air, and wiping with cleanroom-compatible swabs and wipes and appropriate solutions.
- Personnel must wear approved cleanroom garments and hairnets. Feet should be entirely enclosed and the face covered across the mouth. Gloves are required to prevent fingerprints on parts and components.
- Any debris, if not avoidable, should immediately be removed, for example with a vacuum cleaner.
- For shipping, any sensitive parts, for example smooth sealing surfaces (flanges, etc.), should be protected from vibration damage.
- After final assembly and testing, all exposed parts of a robotic system should be thoroughly cleaned with deionized water and cleanroom-approved wipes, and then enclosed in clean, non-shedding polyethylene wrap.

Example 3.2 presents a real-world product specification for a wafer-handling robot.

Example 3.2: specification and design of a wafer-handling robot

The following is the product specification for a typical atmospheric wafer-handling (AWH) robotic system that will operate in an ISO Class 3 cleanroom environment within a 300 mm semiconductor factory. The robotic system includes a robot, a wafer pre-aligner that rotates the wafer to a specified angle, an end-effector with a 'break-the-beam' wafer scanner, and a robot controller.

Purpose and scope

Purpose. The purpose of this product specification is to define the mechanical and electrical hardware, software, and interconnect requirements for the AWH robotic system that includes wafer-handling robot, wafer aligner, and wafer scanner.
Scope. The AWH robotic system will be used for wafer handling in a metrology tool. The specification is limited to this tool only. All requirements comply with the overall tool level requirements.
Negotiable Capabilities. All capabilities listed in this document must be met. Wherever a specific requirement/capability cannot be achieved owing to technology or cost challenges, the supplier will make note and provide explanation.

General description

The basic function of the AWH300 robotic system is to provide the following:

- The robot system removes wafers from cassettes or front-opening unified pods (FOUPs), and orients them as required for the metrology application. After measurement the wafer is placed in the FOUP in the specified wafer slot.
- In order to minimize the wafer swap time (the time needed to replace a processed wafer with a new, unprocessed wafer) a robot with two arms ('dual-arm robot') or two independently controlled end-effectors may be considered.

- The AWH300 robotic system will serve two wafer cassettes arranged at a 90° angle. It is preferred that the robot can handle wafers to/from both cassettes without being mounted to a linear track (an additional linear axis of motion that moves the entire robot).
- A non-contact laser sensor that is able to detect wafers, potential 'cross slots,' and 'double slots' (two wafers in one cassette slot) is required. Any suitable wafer scanner may be considered.
- The robot system pre-aligns the wafer according to the specified wafer target *x*/*y* position and orientation angle.
- The robot system picks and places the wafer from/to the chuck in the metrology tool.
- The robot system automatically detects the size of 200 mm and 300 mm wafers. No hardware changes shall be required for switching between 200 mm and 300 mm wafers.

Cost targets

The target cost is challenging but must nonetheless be met. In cases where robot performance may be compromised because of cost the supplier is invited to propose alternatives.

Wafer-handling parameters

Wafer characteristics. The AWH300 robotic system will be able to handle wafers with the following characteristics:

Wafer dimensions: wafer diameters 200 mm and 300 mm according to the SEMI standard M1. The acceptable wafer thickness ranges are listed in Table 3.11.

Wafer cassettes: as described in Table 3.12. The number of wafers per cassette and the pitch between wafer slots can be specified by the user.

The cassettes comply with SEMI standards E1.9 and E47.1.

Manufacturing environment

Facilities. The AWH robotic system requires the typical facilities specified below.

Power: peak electrical power consumption does not exceed 1250 W for more than 0.3 s. The maximum continuous power consumption does not exceed 650 W. The system is capable of interfacing with the following voltage ranges:

220 VAC ± 10%, 50 Hz
120 VAC ± 10%, 60 Hz

Voltage: voltage auto ranging is preferred.

EMO (emergency off): the robotic system does not cause damage to the wafer in case of a power failure. Once power is restored the system can resume operation with the wafer.

Vacuum: the robotic system requires a vacuum flow rate of no more than $3.0 \cdot 10^{-4}\ m^3 \cdot s^{-1}$ at 67 800 Pa during wafer handling. Motion without a wafer does not need vacuum flow.

Heat dissipation: the robotic system does not require external cooling. Ambient air may be used by the system if the airflow is directed away from the wafer.

Environment. The AWH robotic system operates in an environment with the following parameters.

Relative humidity: 40 to 90% (non-condensing)

Operating temperature: 10 to 40 °C

Shipping and storage temperature: –25 to 55 °C

Altitude (max.): 1 100 m

Environmental cleanliness: ISO Class 3 or better per ISO standard 14644

Dimensions and work envelope. The physical dimensions and the work envelope of the AWH robotic system are compatible with the following specifications.

Z-stroke: the robot has a vertical *Z*-stroke of at least 0.4 m.

Arm reach: the robot has a radial arm reach of at least 0.8 m, including the end-effector.

Theta rotation: the robot has a Θ-rotation of at least 345°.

Sweep radius: the robot can sweep through its full theta rotation range inside a circle with a 0.3 m radius (0.6 m diameter). This sweep radius includes a 300 mm wafer and the cassette mapper.

Physical characteristics and performance

Mechanical parameters.

Arm and end-effector stiffness: the robotic system is capable of processing a payload of at least 0.5 kg. With this payload the maximum permissible deflection of the arm and end-effector at the wafer center is 0.5 mm. The system accommodates a maximum payload of 1.0 kg without sustaining permanent damage.

Mechanical end-effector interface: a suitable mechanical interface for the end-effector, in particular the bolt pattern, must be agreed upon with the supplier.

Weight: the weight of the AWH robotic system is no greater than 70 kg. No single component shall weigh more than 40 kg.

Regulatory compliance. The AWH robotic system will conform to the following EMC and safety standards: SEMI M1-1105, SEMI S2-0703, SEMI S8-0705, SEMI E1.9-0701, SEMI E47.1-0306, IEC 60825-1, UL 2011, and the European Community (CE Mark) Directives LVD 73/23/EEC, MD 89/392/EEC, and EMC 89/336/EEC.

Serviceability, maintenance, and reliability.

Serviceability: all major components of the AWH robotic system are accessible from the exposed areas. The robot does not require removal from the production tool for servicing.

Maintenance: the AWH robotic system requires scheduled maintenance no more than once per year. The scheduled maintenance can be performed in the field and does not take longer than 2.5 h. The service is limited to minor part replacement, adjustments, and lubrication as needed.

Reliability: the mean time between failures (MTBF) of the AWH robotic system is 45 000 h or better, per SEMI standard E10-0304.

Reliability: the mean time between assists (MTBA) of the AWH robotic system is 30 000 h or better, per standard SEMI E10-0304.

Product life: the product life of the AWH robotic system shall be 10 years or better.

Wafer throughput. The AWH robotic system shall achieve the wafer throughput performance listed in Table 3.13. A 300 mm cassette with 25 wafers and a 10 mm pitch between wafers is assumed. The cassette position 1 is the bottom slot. The commanded motion sequences use the worst-case distances per axis: full extension for the R-axis, 180° for the Θ-axis, and full Z-stroke. It is desired to achieve travel times less than the time budget per task listed in Table 3.13. It is assumed that a 'break-the-beam' wafer scanner, embedded in the end-effector, is used for maximum reliability (Figure 3.11). A reflective wafer scanner is acceptable if it operates with a comparable reliability.

Robot positional repeatability. The robot shall place a wafer at the same commanded position with a 3σ positional repeatability of ± 0.13 mm. At a wafer center position (arm extension) of $R = 500$ mm this translates to a Θ-axis positioning repeatability of 0.015° or better. The 3σ positional repeatability for the Z-axis shall be 0.1 mm.

Wafer scanner. The wafer scanner shall use a non-contact sensor. Any laser-type sensors shall conform to IEC standard 60825. The wafer scanner shall be capable of detecting the presence and position of a wafer in the cassettes, or cross-slotted wafers or two wafers in one slot. The detection accuracy for all cases shall be 99.9%.

Teaching mode. The AWH robotic system shall offer the feature of storing ('teaching') at least 20 attained poses (end-effector positions). Taught positions shall not be lost in the event of a power failure.

Environmental control and cleanliness

The AWH robotic system shall operate in Class 3 cleanrooms per ISO 14644–1 without degrading the cleanliness of the environment. This requirement applies to the robot vicinity above the wafer plane during handling.

Mechanical interface

Robot. The robot shall be mountable as specified in a separate drawing and shall be able to operate in the documented environment.

Aligner. The aligner shall be mountable to a flat plate with threaded holes, with cable and hose clearance, as specified in a separate drawing.

End-effector and wafer scanner. If the end-effector and wafer scanner are not integrated with the AWH robotic system, mounting and connection provisions shall be provided that are consistent with the configuration shown in a separate drawing.

Wire harness routing. The connector locations of the AWH robotic system shall be below the mounting plane and routable as shown in a separate drawing.
Vacuum routing. The locations of the vacuum connectors shall be below the mounting plane and routable as shown in a separate drawing. The connectors shall accept tubing with 1.6 mm inner diameter and 3.1 mm outer diameter.

Electrical interface

Communication interface. The system-level interconnection between the metrology tool and the AWH robotic system utilizes either RS232C connections or an Ethernet cable.
Power connector. A standard three-blade IEC, VDE (Verein Deutscher Ingenieure) approved AC connector is preferred.
Communications. Communication among the components within the AWH300 robotic system utilizes a ‘single wire,’ bus-based control system. The specific bus protocol is negotiable, but a commonly used standard is preferred, for example CAN-bus, EtherCAT, Firewire, or similar.

Mechanical robot design

The resulting mechanical robot design is shown in Figures 3.13 and 3.14. Figure 3.13a is a top view of the robot with the arm extended. Note that the arm is not fully extended, that is, the angle between arm links 1 and 2 is less than 180°. This avoids the mathematical singularity at the robot’s workspace boundary, where the arm becomes difficult to control. The laser beam at the end-effector is used as a non-reflective (‘break-the-beam’) wafer scanner. Figure 3.14b shows the robot with the arm in a retracted position that achieves the specified sweep radius for robot rotation in the tool or enclosure. The SCARA-type arm has two independently controlled end-effectors shown in Figure 3.14. This allows for increased throughput. Furthermore, three controlled arm links are needed for serving two or three aligned wafer cassettes without a linear track, for example in an EFEM. □

Table 3.11. Specified wafer thickness.

Wafer size (mm)	Wafer thickness (μm)
200	195 – 775
300	295 – 800

Table 3.12. Wafer cassette specifications.

Wafer size (mm)	Maximum number of wafers per cassette	Pitch between wafer slots (mm)
200	26	6.35 – 10.00
300	25	10.00

Table 3.13. Motion sequence for testing the throughput performance in Example 3.2.

Wafer-handling task	Robot motion	Time budget (s)
Initial robot pose	$R = 0$ mm (end-effector above robot center) $Z = 250$ mm (cassette top slot no. 25) $\Theta = 0°$ (facing the wafer aligner)	0
Scan wafer cassette	Rotate to $\Theta = 180°$ (left 180°), orienting the wafer scanner to the cassette; move to $Z = 0$ mm (down 250 mm); extend arm to $R = 362$ mm (cassette slot 1); move to $Z = 250$ mm while scanning wafers.	5
Pick wafer from cassette position 1	Move to $Z = 0$ mm (down 250 mm); extend the arm to $R = 380$ mm (under the wafer); move to $Z = 5$ mm (pick up wafer); retract arm to $R = 0$ mm.	4
Place wafer on aligner chuck	Rotate to $\Theta = 0°$ (right 180°), orienting the end-effector to aligner; move to $R = 380$ mm (above aligner chuck); move to $Z = 0$ mm (down 5 mm); retract arm to $R = 360$ mm; wait for aligner.	4
Pick wafer from aligner chuck	Extend arm to $R = 380$ mm; move to $Z = 5$ mm (up 5 mm, pick up wafer); retract arm to $R = 0$ mm.	2
Place aligned wafer in cassette	Rotate to $\Theta = 180°$ (left 180°), orienting the wafer scanner to the cassette; extend arm to $R = 380$ mm; move to $Z = 0$ mm (drop wafer); retract arm to $R = 0$ mm.	4

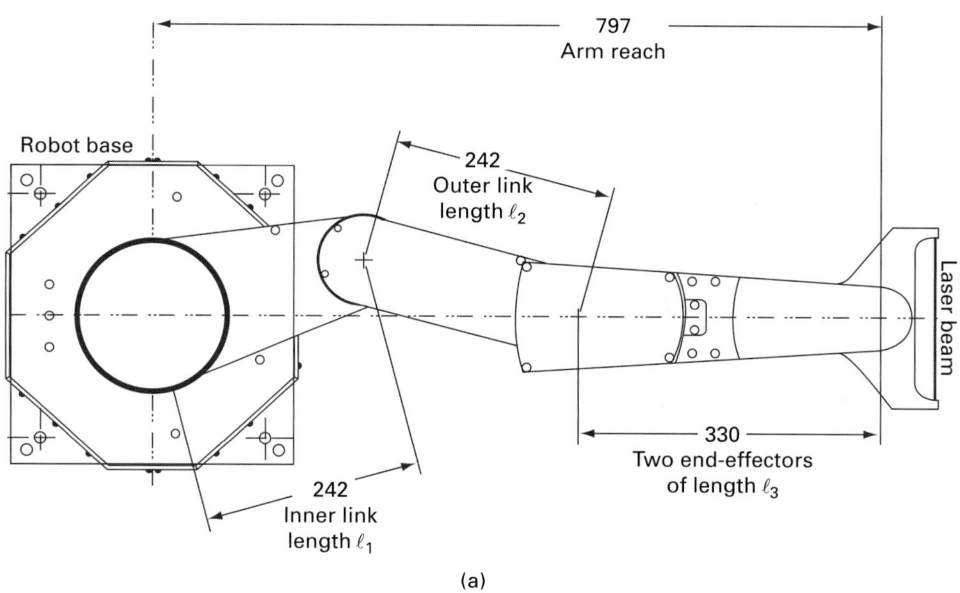

Figure 3.13 Bottom-mounted atmospheric wafer-handling robot in Example 3.2: top view with (a) extended arm, (b) retracted arm. Dimensions are in mm. Source: Cymechs Corp.

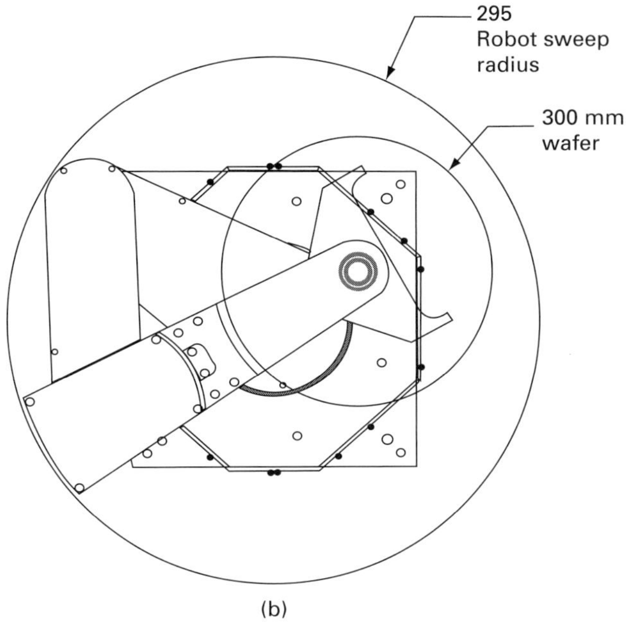

Figure 3.13 (cont.)

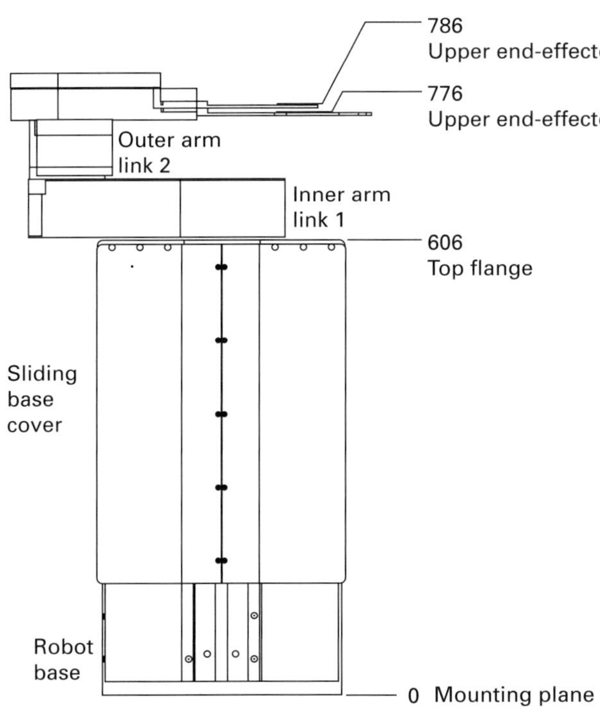

Figure 3.14 Side view of bottom-mounted atmospheric wafer-handling robot in Example 3.2. Dimensions are in mm. Source: Cymechs Corp.

3.8 Applicable and related standards

Several industry standards and guidelines apply directly or indirectly to the robots discussed in this chapter. The following list provides a selection. ANSI standards are published by the American National Standards Institute (ANSI). ASTM standards are published by ASTM International. CE standards are published for the single market in the European Economic Area. ESD standards are published by the Electrostatic Discharge Association (ESDA) and by ANSI. IEC standards are published by the International Electrotechnical Commission (IEC). ISO standards are published by the International Organization for Standardization (ISO). RIA standards are published by the Robotic Industries Association (RIA). Sematech standards are published by International Sematech. SEMI standards are published by Semiconductor Equipment and Materials International (SEMI). UL standards are published by Underwriters Laboratories. Contact information for these organizations is listed in Appendix B.

ANSI/ESD S20.20-2007, Development of an Electrostatic Discharge Control Program for Protection of Electrical and Electronic Parts, Assemblies and Equipment.

ANSI/ESD, S4.1-2006, Work-surface resistance measurements.

ANSI/ESD STM11.11-2006, Surface Resistance Measurement of Static Dissipative Planar Materials.

ANSI/ESD STM11.12-2007, Volume Resistance Measurement of Static Dissipative Planar Materials.

ASTM B117-97, Standard Practice for Operating Salt Spray (Fog) Apparatus, ASTM International, 1997. (Replaced by ASTM B117-02.)

ASTM-D792-00, Standard Test Methods for Density and Specific Gravity (Relative Density) of Plastics by Displacement.

ASTM-D638-03, Standard Test Method for Tensile Properties of Plastics.

ASTM-D256-06a, Standard Test Methods for Determining the Izod Pendulum Impact Resistance of Plastics.

ASTM-D790-03, Standard Test Methods for Flexural Properties of Unreinforced and Reinforced Plastics and Electrical Insulating Materials.

ASTM-D696-03, Standard Test Method for Coefficient of Linear Thermal Expansion of Plastics Between –30°C and 30°C With a Vitreous Silica Dilatometer.

ASTM-D150-98(2004), Standard Test Methods for AC Loss Characteristics and Permittivity (Dielectric Constant) of Solid Electrical Insulation.

ASTM-D257-99(2005), Standard Test Methods for DC Resistance or Conductance of Insulating Materials.

CE 73/23/EEC, European Community (CE Mark), Low Voltage Directive (LVD) 73/23/EEC Amended 93/68/EEC.

CE 89/392/EEC, European Community (CE Mark), Machinery Directive (MD) 89/392/EEC Amended 91/368/EEC, 93/44/EEC, 93/68/EEC.

CE 89/336/EEC, European Community (CE Mark), Electromagnetic Compatibility Directive (EMC) 89/336/EEC Amended 92/31/EEC and 93/68/EEC (also IEC 61000-4-2).

ESD ADV1.0-2004, Glossary of Electrostatic Discharge Terminology.

ESD S6.1-1999, Grounding – Recommended Practice.

ESD TR11-01, Electrostatic Guidelines and Considerations for Cleanrooms and Clean Manufacturing.

ESD TR20.20-2000, ESD Handbook.

ESD STM5.1-1998, Electrostatic Discharge Sensitivity Testing – Human Body Model.

ESD STM5.2-1999, Electrostatic Discharge Sensitivity Testing – Machine Model.

ESD STM5.3.1-1999, Electrostatic Discharge Sensitivity Testing – Charged Device Model.

ANSI/ESD SP10.1-2007, Standard Practice for the Protection of Electrostatic Discharge Susceptible Items – Automated Handling Equipment (AHE).

IEC 60093 Ed. 2.0 b:1980, Methods of test for volume resistivity and surface resistivity of solid electrical insulating materials.

IEC 60250 Ed. 1.0 b:1969, Recommended methods for the determination of the permittivity and dielectric dissipation factor of electrical insulating materials at power, audio and radio frequencies including meter wavelengths.

IEC 60825-1, Safety of laser products – Part 1: Equipment classification and requirements.

IEC 61000-4-2 Ed. 1.2 b:2001, Electromagnetic compatibility (EMC)- Part 4–2: Testing and measurement techniques – Electrostatic discharge immunity test.

International Technology Roadmap for Semiconductors – 2006 Update, Sematech International.

ISO 1302:2002, International Organization for Standardization, Geometrical Product Specifications (GPS) – Indication of surface texture in technical product documentation, Geneva, Switzerland, 2002.

IS0 9283:1998(E), International Organization for Standardization, Manipulating industrial robots – Performance criteria and related test methods, Second edition, Geneva, Switzerland, 1998.

ISO 9787:1999, Manipulating industrial robots – Coordinate systems and motion nomenclatures.

ISO 9946:1999, Manipulating industrial robots – Presentation of characteristics.

ISO 10218-1:2006, Robots for industrial environments – Safety requirements – Part 1: Robot.

ISO 11359-2:1999, Plastics – Thermomechanical analysis (TMA) – Part 2: Determination of coefficient of linear thermal expansion and glass transition temperature.

ISO 14644-1:1999, Cleanrooms and associated controlled environments – Part 1: Classification of air cleanliness. Geneva, Switzerland, 1999.

ISO 14644-2:2000, Cleanrooms and associated controlled environments – Part 2: Specifications for testing and monitoring to prove continued compliance with ISO 14644-1. Geneva, Switzerland, 2000.

ISO 14644-4:2007, Determination of particle size distribution – Single particle light interaction methods – Part 4: Light scattering airborne particle counter for clean spaces. Geneva, Switzerland, 2007.

Sematech (2007), International Technology Roadmap for Semiconductors (ITRS). Austin, Texas.

Sematech 92051107A-STD, International Sematech, Guide to Contamination Control in Design, Assembly, and Delivery of Semiconductor Equipment, Austin, Texas (1992).

Sematech 99033693A-ENG, Integrated Minienvironment Design Best Practices.

SEMI E1.9-0701, Mechanical Specification for Cassettes Used to Transport and Store 300 mm Wafers.

SEMI E10-0304, Specification for Definition and Measurement of Equipment Reliability, Availability, and Maintainability (RAM).

SEMI E14, Measurement of Particle Contamination Contributed to the Product from the Process or Support Tool.

SEMI E21-94, Cluster Tool Module Interface, Mechanical Interface and Wafer Transport Standard.

SEMI E43-0301, Guide for Measuring Static Charge on Objects and Surfaces.

SEMI E47.1-0306, Provisional Mechanical Specification for FOUPS used to Transport and Store 300 mm Wafers.

SEMI E49.6-1103, Guide for subsystem assembly and testing procedures – stainless steel systems.

SEMI E63-1104, Mechanical Specification for 300 mm Box Opener/Loader to Tool Standard (BOLTS-M) Interface.

SEMI E78-0998, Electrostatic compatibility, guide to assess and control electrostatic discharge (ESD) and electrostatic attraction (ESA) for equipment.

SEMI E129-1103, Guide to assess and control electrostatic charge in a semiconductor manufacturing facility.

SEMI F19.0304, Electropolishing Specifications for Semiconductor Applications.

SEMI M1-1105, Specifications for Polished Monocrystalline Silicon Wafers.

SEMI S2-0703, Environmental, Health, and Safety Guideline for Semiconductor Manufacturing Equipment.

SEMI S8-0705, Safety Guidelines for Ergonomic Engineering of Semiconductor Manufacturing Equipment.

UL 2011, Underwriters Laboratories, UL Subject 2011, Factory Automation Equipment.

4 Design of vacuum robots

Certain processes in electronics manufacturing require an ultra-clean vacuum environment or a controlled environment of a gas mixture. This chapter presents engineering guidelines and best practices for the design of material-handling robots in such extreme environments.[1]

4.1 Robotics challenges in vacuum environments

Vacuum pressures used in electronics manufacturing, in particular for semiconductor manufacturing processes, range from low to ultra-high vacuum pressures. While high vacuum enables the contamination-free deposition of materials, ultra-high vacuum (UHV) provides atomically clean substrates. Only UHV preserves atomic-scale clean surfaces for extended time periods of minutes or hours. The design objective for the vacuum robots discussed here and described in Definition 4.1 is to provide the cleanliness and vacuum integrity needed for electronics manufacturing, in particular semiconductor manufacturing, which utilizes the most challenging vacuum environments.

Definition 4.1: A vacuum robot operates under vacuum pressures below the ambient atmospheric pressure. Either the entire robot or a portion of the robot resides in vacuum.

If only a part of the robot operates in vacuum a vacuum barrier, integrated with the robot, separates the vacuum portion from the atmospheric portion. In a typical semiconductor manufacturing tool only the robot arm resides in vacuum, while the robot base with motors and the associated electronics is in atmosphere. This is illustrated in Figure 4.1, where the substrate-handling robot is mounted at the center of a vacuum cluster tool. The robot arm, SCARA-type or a closed kinematic chain ('frog-leg' arm), transfers substrates between load lock and process chamber. Only the end-effector enters the load locks through the narrow openings (Murvihill, 1999). The static and dynamic vacuum barriers indicated in the figure are important parts of the robot design. These barriers are critical features of vacuum robots, as is the vacuum compatibility of the robot's components and materials.

The primary engineering challenges associated with the design of vacuum robots for high and ultra-high vacuum include the following:

[1] This chapter was written with the support of Dr. Martin Peter Aalund.

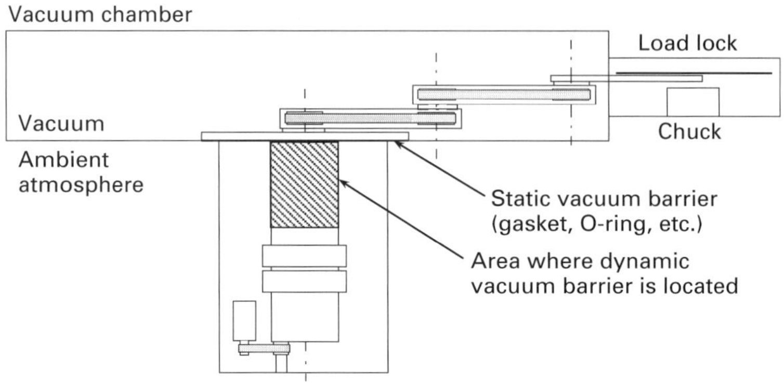

Figure 4.1 Substrate-handling vacuum robot, mounted in a vacuum cluster tool.

- Static vacuum barrier separating the vacuum and atmosphere environments (Section 4.3)
- Dynamic vacuum barrier that transfers motion from atmosphere to vacuum (Section 4.4)
- Clean drive train that does not contaminate the vacuum environment (Section 4.5)
- Prevention of external and virtual leaks that limit the achievable vacuum pressure in the chamber (Section 4.6).
- Surface finishes that prevent outgassing and gas permeation into the vacuum chamber (Section 4.7).

The next section reviews some fundamentals of vacuum science, before design guidelines and best practices for the design of vacuum robots are presented.

4.2 What is vacuum?

4.2.1 Fundamentals

'Vacuum' is broadly defined as space that is relatively empty of matter and where the pressure is below the ambient atmospheric pressure. The transition from atmosphere to vacuum is seamless and without a distinct threshold. Historically the standard atmospheric pressure at sea level was established as a convenient yet arbitrary reference point. This can result in confusing terminology, in which 'ultra-high vacuum' refers to very low pressures, and 'low vacuum' refers to higher pressures close to atmosphere. A physically objective reference point is absolute zero pressure.

The pressure p is defined as the average force F exerted on a surface of unit area A:

$$p = \frac{F}{A}. \tag{4.1}$$

The force is caused by the impact of molecules of a gas or liquid on the walls of the containing chamber. (There are rare exceptions which are not relevant here.) The pascal (Pa) is the SI unit for pressure: 1 Pa = 1 $N{\cdot}m^{-2}$. An officially obsolete but still popular

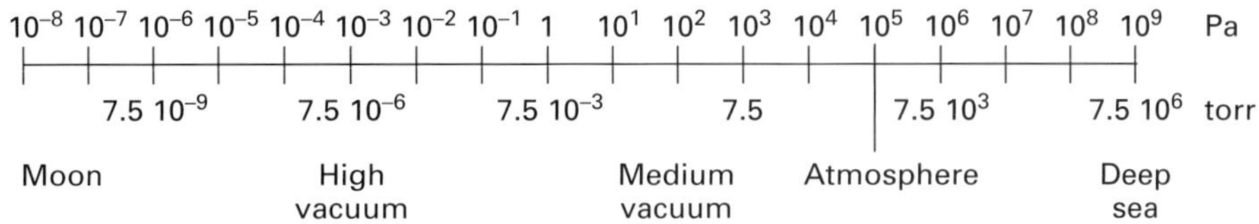

Figure 4.2 Scale of practical pressure ranges.

unit is the torr, or 1 mm of mercury (mm Hg): 1 torr = 133.3 Pa. This is the pressure needed to displace mercury in a manometer by 1 mm and is named after the Italian Evangelista Torricelli, who invented the barometer in 1643. Other obsolete units are 'pound force per square inch' (psi) and 'inches of mercury below atmosphere' (in Hg). The latter 'gauge pressure' uses standard atmosphere as the reference point instead of absolute zero pressure. The absolute vacuum pressure p_{abs} with respect to 0 Pa is the sum of ambient atmospheric pressure p_{atm} = 101 325 Pa = 760 torr and the relative vacuum pressure (or gauge pressure) p_{rel} with respect to p_{atm}:

$$p_{abs} = p_{atm} + p_{rel} = 101\,325 \text{ Pa} + p_{rel} = 29.92 \text{ in} \cdot \text{Hg} + p_{rel}. \tag{4.2}$$

For example, at sea level with $p_{atm} = 101\,325$ Pa the gauge pressure, $p_{rel} = -27$ in Hg is equivalent to $p_{abs} = (29.92 - 27.00)$ in Hg = 2.92 in Hg. Note that in vacuum engineering the minus sign in gauge pressure is usually omitted. More measurement units are listed in Appendix A.

Figure 4.2 shows the practical pressure scale from 10^{-8} to 10^{9} Pa, from Earth's moon to deep-sea ocean trenches. Vacuum ranges from absolute zero (0 Pa) to just below atmospheric pressure. Pressures as low as 10^{-13} Pa are measurable (pressures below that can be estimated). Perfect vacuum (0 Pa) is a largely hypothetical state that cannot be achieved under laboratory conditions. Examples of high-vacuum environments are:

- Cryopumped vacuum chamber: 10^{-7} Pa after 4 h of baking (typical)
- Moon surface: 10^{-10} Pa at night to 10^{-7} Pa in daylight
- Interstellar space: 10^{-16} Pa.

Interstellar space contains only a few molecules per cubic meter. The moon's changing pressure indicates the temperature influence on vacuum environments (R. Bergner, personal communication, 2008).

4.2.2 Vacuum quality

The primary indicator of vacuum quality is the residual pressure, the amount of matter remaining in a pumped-down chamber: less matter indicates a better vacuum level. The pressure ranges in Table 4.1 have been defined by the American Vacuum Society and are often used to categorize vacuum quality (Marquardt, 1999). Low vacuum, or 'rough' vacuum, can be achieved with a vacuum cleaner and can be measured directly with a liquid column manometer. Medium vacuum can be achieved with a mechanical pump and can be measured indirectly with a thermal gauge or a capacitive gauge. High vacuum usually requires multi-stage pumping and cannot be measured but must be inferred from

Table 4.1. Categories of vacuum levels.

Vacuum level	Pressure range (Pa)	Pressure range (torr)
Standard atmospheric pressure	$1.013 \cdot 10^5$	760
Low vacuum	$3.3 \cdot 10^3$ to $1.0 \cdot 10^5$	25 to 760
Medium vacuum	$1.0 \cdot 10^{-1}$ to $3.3 \cdot 10^3$	$7.5 \cdot 10^{-4}$ to 25
High vacuum	$1.0 \cdot 10^{-7}$ to $1.0 \cdot 10^{-1}$	$7.5 \cdot 10^{-10}$ to $7.5 \cdot 10^{-4}$
Ultra-high vacuum (UHV)	$1.0 \cdot 10^{-10}$ to $1.0 \cdot 10^{-7}$	$7.5 \cdot 10^{-13}$ to $7.5 \cdot 10^{-10}$
Extremely high vacuum (EHV)	$< 1.0 \cdot 10^{-10}$	$< 7.5 \cdot 10^{-13}$
Perfect vacuum (absolute zero)	0	0

ion gauge measurements. High vacuum is a significant threshold in process engineering, because the continuum assumptions of fluid mechanics no longer apply (White, 2003). In high vacuum the mean free path (MFP) is longer than the vacuum chamber size. (The MFP is the average distance that a particle travels before it collides with another particle.) In high vacuum at 10^{-4} Pa ($7.5 \cdot 10^{-7}$ torr) a material surface will be fully covered with a gas contaminant after only a few seconds. Ultra-high vacuum (UHV) preserves atomic-scale clean surfaces for longer time periods by reducing the number of molecules in the enclosure that can reach the surface during a given time period. Ultra-high vacuum requires special precautions, including special materials, extreme cleanliness, and baking of the vacuum chamber to remove traces of gases and fluids.

The force $F = p \cdot A$ exerted by vacuum pressures increases only a marginal 0.001% between medium and extremely high vacuum (EHV). Instead of pressure the amount of matter present in the enclosure becomes the indicator of vacuum quality.

4.2.3 Partial pressure

The concept of partial pressure is based on the additive relationship of gas loads. This can be defined as follows.

Definition 4.2: Given a gas mixture in a sealed enclosure, the partial pressure of one gas component is the pressure the gas component would exert if it were alone in the enclosure.

For example, consider a gaseous mixture of oxygen, nitrogen, and water vapor that exerts a total pressure on the enclosure. The partial pressure of oxygen is the pressure developed by the oxygen molecules only. The sum of the partial pressures from all gases equals the total pressure in the enclosure.

4.3 Static vacuum barrier

The static vacuum barrier in Figure 4.1 separates the vacuum environment from the atmospheric environment. The static barrier does not include any moving parts and does not transfer motion from atmosphere to vacuum. The static vacuum barriers for vacuum robots discussed here comprise three main components:

- Two opposing metal flanges
- Bolts or clamps that press the flanges together
- A gasket between the flanges.

The design of these components varies across industries and applications. Specifications for vacuum barriers should not exceed the necessary requirements because systems for higher vacuum are increasingly expensive. The barriers of vacuum robots must withstand the specified maximum pressure differential and also endure the pressure changes between atmosphere and vacuum that result from loading and unloading the process tool. High vacuum demands the vacuum integrity of barriers with an extremely small leak rate. The SI-based unit for the leak rate is $1\ \text{Pa}\cdot\text{m}^3\cdot\text{s}^{-1} = 9.87\ \text{atm}\cdot\text{cm}^3\cdot\text{s}^{-1}$ (Appendix A) and translates as "one cubic meter of gas enters the enclosure per second at a pressure differential of one pascal" (1 atm = 101 325 Pa). The Sematech standard 92051107A-STD recommends that vacuum systems be tested with a helium mass spectrometer leak detector with a sensitivity of $10^{-10}\ \text{atm}\cdot\text{cm}^3\cdot\text{s}^{-1}$ or better. This sensitivity translates to a vacuum integrity of 0.315 cm^3 per 100 years. Bubble testing can detect leaks of down to $10^{-3}\ \text{atm}\cdot\text{cm}^3\cdot\text{s}^{-1}$. Halogen sniffers can detect leaks of $10^{-5}\ \text{atm}\cdot\text{cm}^3\cdot\text{s}^{-1}$.

4.3.1 Flanges

Most vacuum robots are designed for the enormous forces under medium to ultra-high vacuum. For example, a robot mounting flange with an area of $A = 400\ \text{cm}^2$ under a medium vacuum pressure of $p_{\text{abs}} = 1.0$ Pa is exposed to a force of

$$F = p_{\text{rel}} \cdot A = (p_{\text{abs}} - p_{\text{atm}}) \cdot A = (1 - 101\,325)\ \text{Pa} \cdot 0.04\ \text{m}^2 = -4053\ \text{N}. \qquad (4.3)$$

This force is equivalent to a mass of 413 kg and demonstrates that good engineering designs are needed to provide reliable vacuum barriers. Furthermore, the force is not constant, but changes between atmospheric and vacuum during the loading and unloading of substrates.

Standards

Standard, off-the-shelf flanges (or 'couplings') offer a cost-efficient and reliable solution for vacuum robots. These products provide solutions for pressures down to 10^{-11} Pa ($7.5 \cdot 10^{-13}$ torr). Maximum temperatures of up to 500 °C allow 'baking' that removes volatile compounds. Available industry standards and de facto standards that specify dimensions and materials of flanges include:

- ISO standards
- ANSI standards
- Proprietary standard for Conflat® flanges.

ISO standards identify flanges by the approximate diameter of the connected tubing. For example, a KF-50 flange has a specific KF shape and is designed for tubes with a 50 mm diameter. Applicable standards include ISO 1609, ISO 2861, ISO 3669, and ISO/PRF TS

3669–2. The Pneurop standard 6606 is also part of the ISO system. ISO-compliant flanges and 'quick-release' couplings are designed to facilitate assembly procedures. ANSI standards also specify couplings suitable for high vacuum pressures. The Conflat® flange format is designed for high- and ultra-high-vacuum applications and utilizes a copper gasket and a circular bolt pattern with multiple bolts. Some common flange, bolt, and gasket materials are listed in Section 4.7.3.

Flange connections

Several standard flange connections are available. Bolted flange connections pull the flanges together to form a seal between the gasket, or a central O-ring, and both flange faces. Applicable standards are ANSI, ISO-BF, ISO-LF, ISO-NW, as well as the bolted connection for Conflat® flanges. Clamped flange connections use a clamp that fits around both flanges. The following standards specify clamped flanges: ISO-KF, ISO-QF, ISO-Pneurop, ISO-NW, ISO-MF, ISO-PF, ISO-LF, and ISO-K.

Threaded flange connections are available, but are limited to medium vacuum pressures. Virtual leaks (Section 4.6.2) are possible between threads and blind holes, and elastomeric seals used for threaded connections can outgas. However, special threaded connections with venting slots and a tape seal, for example made of Teflon®, can support pressures down to 10^{-4} Pa ($7.5 \cdot 10^{-6}$ torr) and baking temperatures up to 260 °C.

4.3.2 Gaskets and O-rings

The choice of O-ring materials influences the pressure and temperature range of a vacuum seal. Several gasket standards are available, including ISO-BF, ISO-KF, ISO-NW, ISO-QF, ISO-Pneurop, ISO-PF, ISO-LF, ISO-K, ISO-CF, VSR (Swagelok fitting), and the Conflat® format.

Elastomer gaskets or O-rings can be sufficient down to medium vacuum, in some cases to high vacuum, but are not suitable for ultra-high vacuum due to the potential outgassing of volatile compounds. The seal is formed between the central gasket and the two flange faces. One flange has a smooth, flat face, while the opposing flange has a groove for retaining the gasket. An advantage of elastomer seals is re-usability: these materials are not permanently deformed after the first use.

Metal gaskets in the form of flat rings and wire seals are suitable for ultra-high vacuum (UHV) pressures. Virtually all UHV seals are metal gaskets made of soft metals with a low vapor pressure. A prominent example is electrolytic grade copper, called oxygen-free high conductivity (OFHC) copper. Copper ring seals are pressed between the knife-edges of two opposing flanges that cut into the gasket and form a strong vacuum seal (Figure 4.3). As the copper gasket is deformed the 'cold' metal flow fills small surface imperfections in the knife-edge. Such gaskets are specified for vacuum pressures down to 10^{-10} Pa ($7.5 \cdot 10^{-12}$ torr) and for temperatures up to 450 °C. Metal gaskets greatly reduce the risk of permeability and outgassing compared to elastomer gaskets. During system assembly it is important to avoid stress in the seal by evenly tightening the bolts on the bolt circle. A torque wrench is recommended for incrementally increasing the torque (Westerberg, 1999). A disadvantage of metal gaskets is the single-use limitation resulting

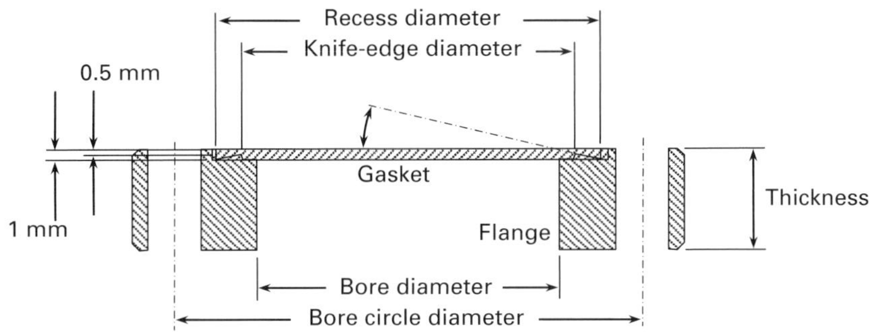

Figure 4.3 Cross section of a Conflat® flange.

from the deformation caused by the connection. A second reliable seal cannot be formed because of deep grooves in the gasket. For repeated use an elastomer material is recommended, although at the cost of a reduced vacuum pressure.

Typical specifications for the material, vacuum range, and temperature range of elastomer and metal vacuum seals are given below.

- Material:
  Flange: stainless steel (304, 304L, 316LN, 316LN ESR), aluminum
  Gasket: OFE Copper, Viton® elastomer
  Bolts: stainless steel 300, 18–8
- Maximum vacuum pressure:
  Wire or CF gasket: 10^{-9} to 10^{-10} Pa ($7.5 \cdot 10^{-12}$ to $7.5 \cdot 10^{-13}$ torr)
  Elastomer seal: 10^{-5} Pa ($7.5 \cdot 10^{-8}$ torr)
- Temperature range:
  Wire or CF gasket: –200 to 450 °C
  Elastomer seal: –20 to 150 °C, intermittent 200 °C

External leaks at a flange can be caused by gaps in the knife-edge that are too large to be filled by the metal copper gasket. Gaskets with circular scratches sometimes maintain ultra-high vacuum, but radial scratches usually cause external leaks across the seal and require gasket replacement. This emphasizes the importance of suitable assembly procedures under cleanroom conditions, and of careful handling and packaging, in order to prevent external leaks (see Section 4.8.2).

4.4 Dynamic vacuum barrier

Vacuum robots in electronics manufacturing operate only partially in vacuum: the robot arm is inside a vacuum chamber, while the robot base resides in a cleanroom environment under ambient pressure. Motion is generated in the robot base by actuators (motors) and transferred across a dynamic vacuum barrier into vacuum to drive the arm (Figure 4.1). This requires a suitable dynamic vacuum barrier that is integrated with the robot. It is generally recommended that a commercially available dynamic seal be selected.

The barrier must provide the necessary vacuum integrity under the specified vacuum pressure. Furthermore, it must not significantly reduce the transferred energy or torque. The available technologies are:

- Magnetic feedthroughs
- Bellows
- Magnetic couplings
- Motors with integrated vacuum barriers
- Lip seals
- Harmonic drives.

The cost of a dynamic vacuum barrier can be significant, depending on the specified vacuum pressure and the number of axes of motion to be transferred into vacuum.

4.4.1 Magnetic feedthroughs

Magnetic feedthroughs are hermetic, direct-drive rotary seals that are commercially available for revolute joints. The direct drive allows for high-torque and high-speed applications. Magnetic feedthroughs are frequently used for vacuum robotics, due to their low contamination, zero backlash, and long product life (Ferrotec, 2001; Rigaku, 2005).

Concept and main components

The main components of the rotating seal are the ferrofluid, the rotating shaft, two stationary pole pieces, and a permanent magnet surrounding the shaft (Figure 4.4). This technical concept has been implemented in various configurations. The ferrofluid is a suspension of ferromagnetic particles in a carrier liquid. Low viscosity and molecular interaction prevent the dispersed particles from settling. The particles, with an average size of 10 to 100 nm, can also be coated with a material that prevents particle agglomeration even in the presence of magnetic fields. As a result, ferrofluids can be polarized and formed with a magnetic field, although they are not ferromagnetic and do not retain magnetization in the absence of an external magnetic field. This property is utilized in magnetic feedthroughs: the magnetic flux across the narrow gaps between the rotating shaft and the pole pieces exerts a force on the ferrofluid that holds the particles, and therefore the entire fluid, firmly in place. The ferrofluid forms a stationary fluid ring seal, a vacuum barrier that surrounds the rotating shaft. The magnetic flux, provided by a permanent magnet, creates a magnetic circuit that includes the pole pieces and the magnetically permeable shaft. The lack of solid friction results in low maintenance requirements and a typically long product life.

The vacuum operating range of magnetic feedthroughs is determined by several design and environmental parameters, most importantly the

- Magnetic design
- Ferrofluid
- Pressure differential
- Ambient temperature.

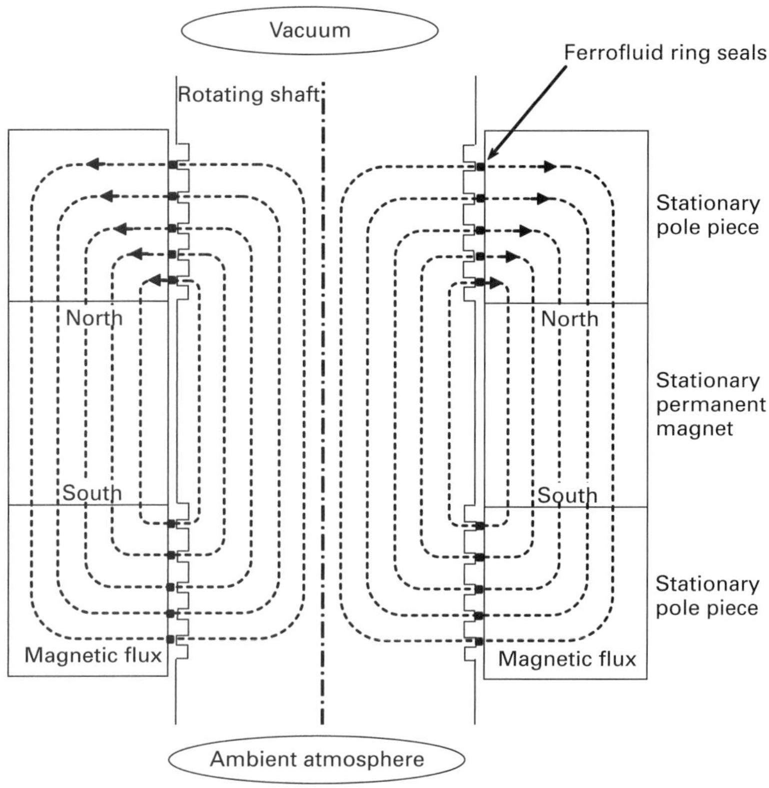

Figure 4.4 Concept and main components of a magnetic feedthrough.

For example, one magnetic feedthrough manufacturer tested a feedthrough using a helium leak test and measured a leak rate (vacuum integrity) of 10^{-7} atm·cm^3·s^{-1} (Ferrotec, 2001; Helgeland, 2002).

Magnetic design

Precise dimensions and locations of shaft and pole tips are critical for the magnetic design. Decreasing gap widths allow increasing pressure differentials. Typical gap widths are about 100 μm. Individual ferrofluid ring seals can withstand pressure differentials of 20 000 to 100 000 Pa (150 to 750 torr) (Ferrotec, 2001; Rigaku, 2005). Higher pressure differentials can be achieved with multiple ring seals, including high-vacuum. The total specified pressure for a magnetic feedthrough is the sum of the admissible pressures of individual seals. For example, if the desired base pressure for a vacuum chamber is 10^{-6} Pa ($7.5 \cdot 10^{-9}$ torr), the pressure differential across the seal is

$$(10^5 - 10^{-6})\,\text{Pa} \approx 10^5\,\text{Pa}. \tag{4.4}$$

The minimum number of required individual seals is

$$N = \frac{100\,000\ \text{Pa}}{20\,000\ \text{Pa}} = 5. \tag{4.5}$$

Ferrofluid

The first fluid ring seal is exposed to the manufacturing process, either a vacuum or (possibly aggressive) gaseous environment. At a vacuum pressure below the ferrofluid's vapor pressure the carrier liquid will increasingly evaporate into the vacuum chamber at a rate approximately proportional to the vapor pressure. The vapor pressure depends on the particular fluid and its temperature. Two carrier fluids and their approximate vapor pressures p_V at 20 °C are:

- Hydrocarbons: $p_V = 10^{-6}$ Pa $(7.5 \cdot 10^{-9}$ torr$)$
- Perfluorinated polyethers (PFPE): $p_V = 10^{-10}$ Pa $(7.5 \cdot 10^{-13}$ torr$)$.

See Section 4.7.1 for more details about vapor pressure. In addition to vacuum and vapor pressures, the following process parameters can affect the ferrofluid and thus the feedthrough's operating range:

- Aggressive and reactive process environments can be harmful to ferrofluids based on hydrocarbons.
- Particles generated by the process can penetrate the fluid seal, enter the narrow gap, and impact the seal quality.
- Some solvents used for cleaning and leak search techniques can destabilize the ferrofluid.

Fluorocarbon-based ferrofluids are often used in aggressive and reactive environments. Vacuum-compatible PFPE grease is also suited for lubricating the bearing that is exposed to the process.

Operating range

The operating range of a typical magnetic feedthrough as a function of the pressure differential and the feedthrough's temperature is shown in Figure 4.5. Assuming an otherwise constant feedthrough configuration, the ferrofluid has a predominant impact on the operating range because of its temperature-dependent vapor pressure. At temperatures below 10 °C heating may be required, to limit friction losses resulting from the

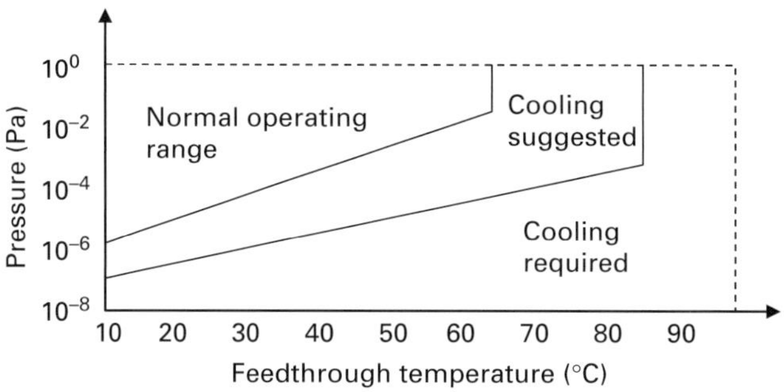

Figure 4.5 Operating range of ferrofluids in vacuum, after Helgeland (2002).

reduced viscosity of the ferrofluid. At very high pressure differentials (at ultra-high vacuum), increased vaporization of the ferrofluid can be limited with cooling.

Example 4.1 presents a dynamic vacuum barrier with two concentric shafts and feedthroughs for a two-link planar robot arm.

Example 4.1: magnetic feedthrough for two-link planar robot arm

Several magnetic feedthrough solutions exist for vacuum robots with one or more axes of motion in vacuum. This example presents a common dynamic vacuum barrier with two rotating shafts for driving a planar, two-link robot arm. Such a barrier can utilize two nested magnetic feedthroughs with two concentric, rotating shafts (Figure 4.6). The inner shaft is supported by a pair of 'inner bearings' near the top and bottom, while the outer shaft is supported by a pair of 'outer bearings.' The two shafts can be independently controlled in atmosphere to drive the two arm links across the barrier in vacuum. Typical materials used for magnetic feedthroughs are listed in Table 4.2, which compiles data from several manufacturers. □

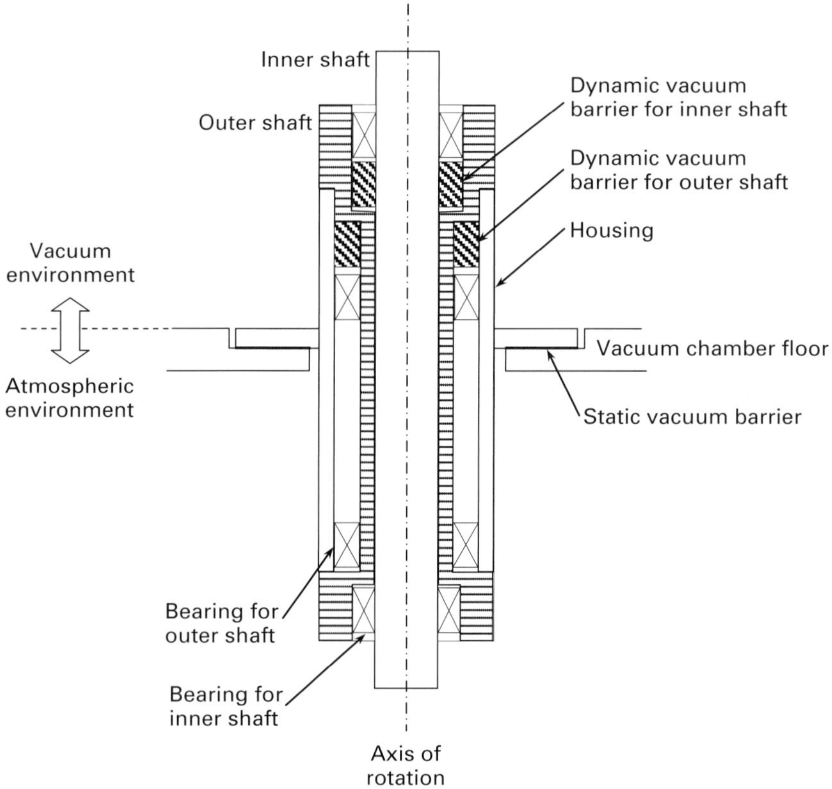

Figure 4.6 Magnetic feedthrough with concentric shafts for two axes of motion.

Table 4.2. Sample materials for the dynamic vacuum barrier in Figure 4.6.

Part or component	Material
Housing, flange	Stainless steel 300-SS5, 303 or 304, not magnetic
Shafts, pole pieces, seal housing	Stainless steel, 400-SS6, 416, 17–4PH
Bearings	Balls and rings: bearing steel, SAE 52100 (SUJ 2), 440C SST
	Retainers: bearing steel, carbon strip steel, stainless steel, PEEK
Magnets	Magnet alloys, SmCo, NdFeB, AlNiCo, isolated from vacuum or process
O-rings, seals (static)	Elastomers (Viton® and others)
Ferrofluid	Hydrocarbons, fluorocarbons (details are proprietary)
	Base: alkylnaphthalene, perfluorinated polyether

4.4.2 Metal bellows

Metal bellows are used to seal prismatic joints of vacuum robots for the transfer of linear motion from atmosphere to vacuum. Metal bellows represent a modern technology with little resemblance to their ancient roots, the leather bellows used in fireplaces and forges for centuries. The predominant bellows types are (a) formed and (b) welded, or 'edge-welded', metal bellows. Formed metal bellows are produced by mechanical forming and have a higher tooling cost but lower manufacturing costs than welded metal bellows. Welded metal bellows consist of metal discs, with 'diaphragms' welded to the two adjacent discs of inner and outer diameter, respectively. A cross section of a welded bellows is shown in Figure 4.7. Two welded diaphragms form one 'convolution.' A group of several convolutions is sometimes called a 'section.' Off-the-shelf sections of a specified length can be combined to a desired length. Welded bellows offer long strokes and fairly precise spring constants. However, the welding of diaphragms is a precision process, typically performed under laboratory conditions with a high visual magnification of the welding joint. This reduces the tooling costs, but increases the manufacturing costs. Welded metal bellows are often the preferred choice in vacuum robotics.

Metal bellows are manufactured from a variety of high-strength metals and alloys, such as stainless steel and titanium. Typical specifications are:

- Vacuum pressure down to 10^{-7} Pa ($7.5 \cdot 10^{-10}$ torr)
- Temperature range of several hundred degrees Celsius
- Chemical and corrosion resistance to many aggressive liquids and gases
- Mechanical sensitivity to some damage, such as nicks or dents.

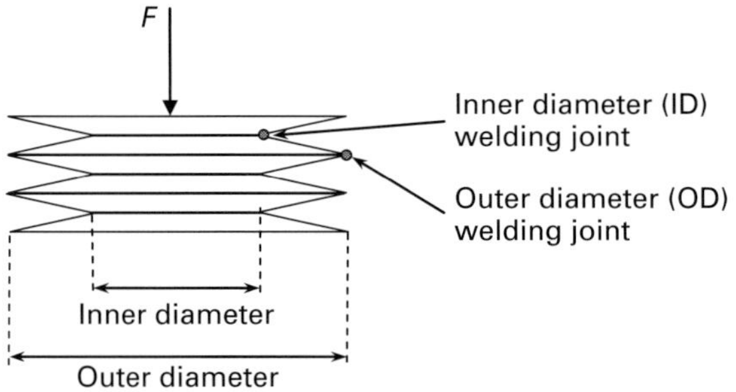

Figure 4.7 Cross section of a welded metal bellows.

An advantage of metal bellows is their compactness (maximum extension relative to minimum compression). Metal bellows offer a generous motion range for the linear dynamic vacuum barriers of vacuum robots and are often used for sealing the vertical axis of motion, with motion ranges from 25 to 75 mm.

During the robot design process, the following bellows parameters should be specified for the manufacturer:

- Free length: length of the unloaded bellows (longitudinal dimension)
- Minimum closed length: minimum possible compressed length
- Stroke: length of the axial motion, either an extension or a compression
- Effective bellows area A_{eff}: used for computing the axial force F exerted by the pressure $p = F/A_{eff}$
- Effective diameter d_{eff}: used to define the effective area $A_{eff} = \pi \cdot (d_{eff}/2)^2$. This is different from the mean diameter d_{mean}, defined by the inner and outer diameters: $d_{mean} = (d_{in} + d_{out})/2$
- Pitch: distance between two adjacent weld beads at the outer edges
- Spring rate: constant force ratio that compresses a bellows by a certain distance
- Volume V: the bellows volume is defined as effective area multiplied by the length L: $V = A_{eff} \cdot L$
- Inner and outer bellows diameters (Figure 4.7).

The product life of a bellows is measured in 'cycles,' hence the term 'cycle life.' Each cycle consists of one extension and one compression. If a metal bellows is used, its cycle life contributes to the overall reliability of a vacuum robot. In electronics manufacturing, robot reliability is specified by the number of mean cycles between failures (MCBF). Tool manufacturers and factory managers expect a robot reliability of several million MCBF. The cycle life of a bellows is influenced by several parameters, including the bellows material, wall thickness, operational stroke, and the number of diaphragms. During operation the specified cycle life of welded metal bellows can be reached or exceeded with the following basic guidelines:

- The robot motion range should not exceed the bellows' free length. The free length can be increased by adding diaphragms.
- Operational strokes should only be compressions from the free length. Repeated extensions beyond the free length will cause material deformation and stress, and ultimately result in a reduced cycle life.
- Welded metal bellows generally withstand higher internal than external pressures.
- For harsh environments the appropriate bellows material must be selected.

These guidelines should be considered when designing or improving a vacuum robot.

4.4.3 Magnetic couplings

Magnetic couplings are hermetic seals suitable for both revolute and prismatic joints. They are used in several commercial vacuum robots and consist of two magnet assemblies, the 'driver' assembly and the 'driven' assembly. A series of radially oriented individual magnets is attached to the magnet assemblies with a high-performance epoxy. The outer driver assembly is connected to a rotating motor shaft in atmosphere

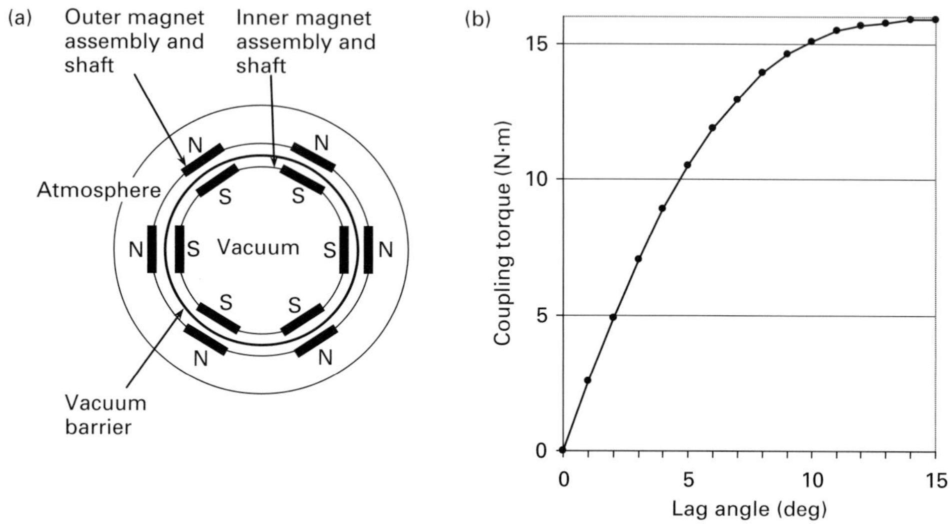

Figure 4.8 (a) Rotary magnetic coupling with permanent magnets; (b) coupling torque as a function of the lag angle, after Kanetomo et al. (1997).

and drives the inner 'driven' assembly and shaft, which reside in vacuum and are attached to the robot arm. The magnet assemblies are separated by a simple wall barrier. There is no physical contact, so the risk of leaks is eliminated.

The concept of a rotary magnetic coupling is shown in Figure 4.8a. It comprises two concentric and coplanar magnet assemblies that couple the inner, driven shaft across a vacuum barrier, which is concentric with the shafts. The inner assembly is in vacuum, driven by the coupling force created by the opposing permanent magnets. The design parameters (barrier thickness, gap between magnet assemblies, magnet type) can be adjusted to achieve a sufficiently stiff force for a given application. When a motor torque is applied, the two magnet assemblies deflect angularly and the (inner) driven magnets lag behind the (outer) driver magnets. This 'lag angle' increases the magnetic force between the two magnet assemblies. The force transfers the torque across the vacuum barrier from the motor to the drive shaft. The coupling torque is, by design, larger than the load torque. Stiff magnetic couplings that allow only small lag angles are created for accurate motion control. The 'softness' of the non-contact coupling is usually small and can be compensated for with a motion-control algorithm. Figure 4.8b illustrates the relationship between the lag angle and the coupling torque. The test data is from an assembly with magnet pairs every 30°, resulting in torque maxima every 30°, that is, in the middle between two adjacent magnet pairs (Kanetomo et al., 1997). At 0° deflection the coupling torque is minimal. A positional repeatability of 0.2 mm was achieved over 125 000 test cycles by the vacuum robot. A coupling torque of about 2.94 N·m was measured at 1°, and of 9.81 N·m at a 5° deflection angle. The internal vacuum pressure was 10^{-7} Pa, and only a few particles larger than 0.1 μm were detected near the robot arm. This performance demonstrates the suitability of magnetic couplings for vacuum robots. In fact, some commercial substrate-handling robots utilize this type of dynamic vacuum barrier.

Example 4.2: magnetic coupling for vacuum indexer

The magnet assemblies of linear magnetic couplings are guided by linear rails. The use of a magnetic coupling for a prismatic joint is shown in Figure 4.9. The joint is part of a vacuum 'indexer' that is mounted to a load lock and vacuum chamber. Indexers are needed because the size of vacuum chambers is as small as practical in order to minimize the pumpdown time. This in turn reduces the vertical motion range of vacuum robots to a few centimeters, just sufficient for loading and unloading ('picking and placing') substrates at locations with the same vertical position. Instead of elevating the robot arm and end-effector, as is the case with atmospheric robots, the substrate carrier or wafer cassette is elevated by the indexer to a desired slot position. The substrates can then be loaded or unloaded from that position by the vacuum robot.

The driving magnet assembly moves vertically while the inner magnet assembly follows closely. In Figure 4.9 the outer driver magnet assembly is controlled by a motor, while the inner magnet assembly elevates the wafer cassette to the commanded position. A load lock interfaces the indexer and the chamber. The magnets in vacuum can be protected from aggressive gases with non-magnetic stainless steel or aluminum enclosures, or with Teflon®, electroless nickel plating, or other suitable material. For example, uncoated neodymium rare earth magnets would rust rapidly in a hydrogen environment. □

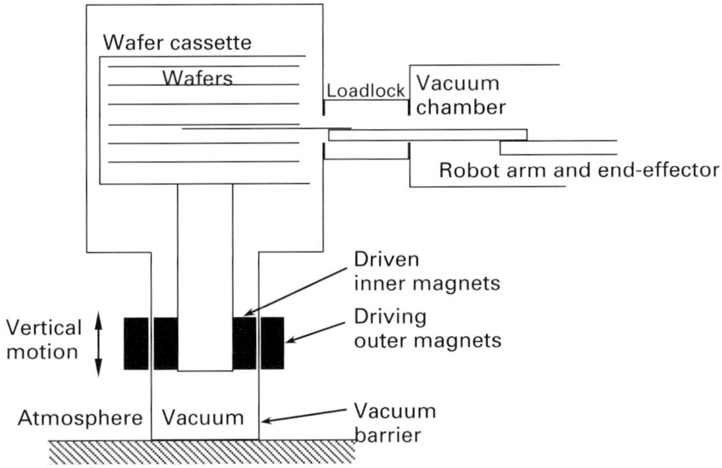

Figure 4.9 The linear magnetic coupling actuates a wafer cassette indexer in vacuum. The vacuum chamber and robot are partially shown.

The design objective is a sufficiently large, fairly constant coupling torque. A second objective may be small dynamic seal dimensions. This requires an application-dependent number of magnet pairs of a certain size. The coupling torque is a function of both the lag angle and the magnetic flux density between the inner and outer magnets. This suggests that the seal dimensions can be minimized by:

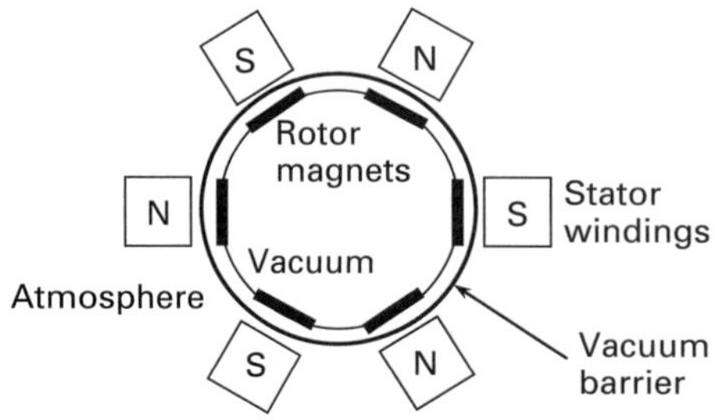

Figure 4.10 Permanent magnet motor with vacuum barrier, permanent rotor magnets, and stator windings.

- minimizing the gap between the inner and outer magnet assemblies. This requires minimizing the barrier thickness and the use of a suitable non-magnetic material.
- selecting a small but sufficient source for the magnetic flux. Permanent magnets generate a considerable magnetic energy relative to their size. Rare earth magnets are excellent, but expensive.

Rotary magnetic couplings are used for vacuum barriers in revolute joints. Linear magnetic couplings are also available and are used for prismatic joints (Example 4.2).

4.4.4 Motors with integrated vacuum barriers

Special electrical motor designs with an integrated vacuum barrier between rotor and stator offer rotary dynamic seals suitable for vacuum robots. Commercial substrate-handling robots are available that utilize this concept for controlling a robot arm in vacuum. This concept is similar to the magnetic coupling in Figure 4.8. The difference is that the stator windings replace the permanent magnets. The rotor utilizes permanent magnets (Figure 4.10). An integrated vacuum barrier between stator and rotor is possible for various motor types, including inductance and reluctance motors. The advantages of this approach include good vacuum isolation and a simple yet efficient design. A disadvantage is the presence of bearings in vacuum, which pose reliability and contamination risks. The reduced magnetic flux across the vacuum barrier, and therefore a reduced coupling torque, can impact both motor stiffness and precision control. However, test data (Kanetomo et al., 1997), as well as the successful performance of robots installed in the field, demonstrate the viability of motors with integrated vacuum barriers. For example, two motors can be vertically stacked into a multi-axis vacuum motor assembly for controlling a SCARA-type arm as described in Section 3. Each motor would drive one axis of motion inside the vacuum chamber.

4.4.5 Lip seals

Lip seals are widely used in the food, pharmaceutical, petrochemical, and semiconductor industries. They are available in various sizes and materials, mostly polymers, and are used under dry conditions without lubrication. Many lip seal products are extruded and cut, and

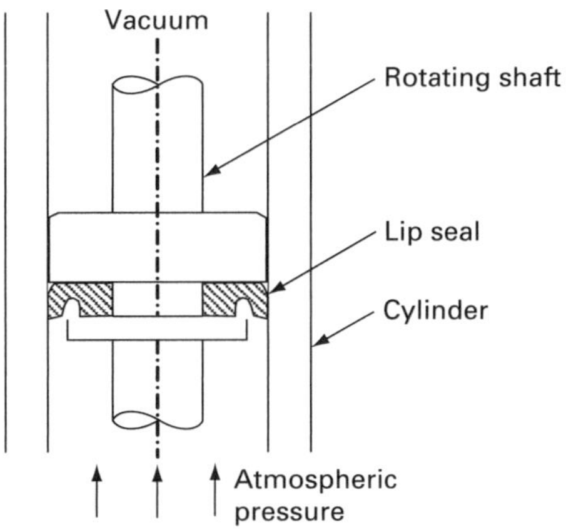

Figure 4.11 Lip seal as a dynamic vacuum barrier for linear or rotary motion.

Table 4.3. Estimated vacuum level and market share of semiconductor manufacturing processes. Source: Cymechs Corp.

Process	Estimated market share	Typical vacuum level
Molecular beam epitaxy	3%	Ultra-high
E-beam, sputtering, physical vapor deposition	11%	Ultra-high to high
Ion implant	11%	High
Inspection and metrology	9%	High
Ashing	3%	High to low
Etch	28%	Medium
Atomic layer deposition	1%	Medium
Chemical vapor deposition	30%	Medium to low

are therefore relatively inexpensive. Lip seals produce low, constant friction over a wide pressure range. Some special designs have a geometry that adjusts the sealing force to the applied pressure, therefore reducing the friction to a desired value. This feature is limited to applications where pressure is being exerted in one direction only, as is the case with the vacuum robots discussed here. Lip seals are suitable for both linear and rotary dynamic seals. Nested concentric shafts as discussed earlier are also feasible with lip seals.

Lip seals can be used for vacuum robots as either static or dynamic seals, in low and medium vacuum environments with pressures down to approximately 1 Pa ($7.5 \cdot 10^{-3}$ torr). The rotary lip seal in Figure 4.11 is also suitable for prismatic joints, for example the vertical axis of motion of substrate-handling vacuum robots. However, despite their low cost and technical suitability lip seals are rarely used in vacuum robotics, although the estimated market share of low and medium vacuum applications is between 30% and 60%, depending on the application (Table 4.3). Only about 35% of semiconductor manufacturing processes require high to ultra-high vacuum. This seems to justify cost-efficient robot products dedicated to moderate vacuum pressures only. Note that process recipes are closely guarded business secrets, so the values in Table 4.3 were estimated using secondary indicators (K. Park, personal communication, 2009). One might speculate that perceived reliability and cleanliness concerns are one reason, as well as the perception that a robot product should serve the larger market from low- to ultra-high-vacuum applications.

4.4.6 Harmonic drives

A harmonic drive is a gear reduction mechanism that offers high gear ratios in a small volume. The concept was introduced in 1957 (Musser, 1960) and has been used for some atmospheric wafer-handling robots since 1986. A harmonic drive can also be used, with the necessary modifications, for vacuum robots as a dynamic vacuum barrier, a gearbox with an integrated vacuum barrier. It is believed that this concept has not yet been implemented and tested. (Hermetically sealed harmonic drives were developed for space use in 1965.) The three basic components of a harmonic drive are the wave generator, the flexspline, and the circular spline (Figure 4.12). The flexspline's rotating steel cup, properly designed, can provide the proposed vacuum barrier. One harmonic drive would be needed per axis of motion in vacuum.

The input to the harmonic drive is the motor torque, and the output is the flexspline torque. In the normal configuration shown in the figure the input torque is applied to the wave generator, a steel disc with an elliptical shape. The rotating wave generator drives the flexspline, a steel cup with a thin wall that is radially flexible yet torsionally stiff. The gear teeth are machined into the outer surface on the vacuum side. This gear interacts with the gear attached to the circular spline, a thick-walled, rigid ring that is the fixed, non-rotating part of the harmonic drive.

The wave generator is inserted in the flexspline. The rotating wave generator applies a continuously moving ellipse to the flexspline. The wave-like motion results in a periodic engaging and disengaging between the external flexspline teeth and the internal teeth of

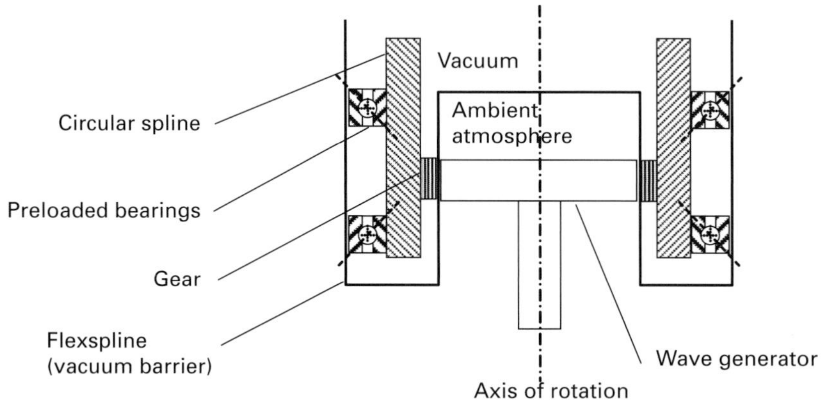

Figure 4.12 Cross section of a harmonic drive with vacuum barrier.

the circular spline. The flexspline has fewer teeth than the circular spline, each revolution shifting the two opposing points of engagement by a few teeth. The gear ratio of the harmonic drive is determined by the flexspline's small output displacement per full revolution of the wave generator input: the gear ratio is the number of teeth on the flexspline divided by the difference between the number of teeth on the circular spline and on the flexspline.

For example, a circular spline with $N_c = 202$ teeth and a flexspline with $N_f = 200$ teeth implement a gear ratio R of

$$R = \frac{N_f}{N_c - N_f} = \frac{200}{202 - 200} = 100. \tag{4.6}$$

Ratios on the order of 100:1 are possible in a small space where conventional planetary gears achieve ratios of only 10:1. The flexspline rotation is always in the opposite direction to the wave generator's rotation. Another advantage of harmonic drives, besides high gear ratios in a small package, is the zero backlash transmission. Disadvantages include a slight elasticity and a torque limit for the transmission.

4.4.7 Motors and electrical components in vacuum?

Electrical parts and components should not be used in high and ultra-high vacuum for several reasons:

- Wiring: solder connections and wire insulation can outgas in high and ultra-high vacuum.
- Vacuum: standard motors are not specified for vacuum pressures below approximately 10^{-2} Pa, owing to the presence of bearing grease, paper slot liners, coatings, solder, and winding insulation. This poses a contamination risk due to particle generation and outgassing.
- Vapor pressure: many lubricants have a vapor pressure above the base pressure for a manufacturing process and therefore would vaporize and pose a contamination risk.
- Cooling: the heat produced by electric components must be removed from the hermetic vacuum chamber using heat sinks, heat pipes, radiation, or other means. This increases the complexity and cost of a vacuum system.
- Electrical wiring: power and signal wiring must exit a vacuum chamber through static seals. This would increase the complexity and cost of a vacuum system. Air trapped in insulation can create virtual leaks, or rupture the insulation and cause particle contamination.

The above emphasizes the technical risks of using electrical parts and components in high and ultra-high vacuum. However, technical solutions are available for motors and the associated electronics in low and medium vacuum. It is recommended that only vacuum-compatible materials and parts, electrical components coated with vacuum-compatible compounds and epoxy, and 'through-the-arm' wiring for sensors and actuators integrated with the end-effector be used (Aalund, 1998).

4.5 Clean drive trains

A vacuum robot must not allow contaminants to enter the vacuum chamber from the ambient environment through external leaks, and it must minimize the generation of contaminants inside a vacuum chamber. Both contamination sources could compromise the required cleanliness and vacuum pressure. The drive train of a robot arm comprises several moving parts and therefore poses a contamination risk that can be mitigated using best practices and guidelines for vacuum robots.

4.5.1 Transmission with metal belts and pulleys

Metal belts are used for vacuum robots instead of the plastic and composite belts common in atmospheric robots. Metal belts are made of various metal alloys, depending on the desired properties, and have several advantages for vacuum applications:

- Negligible particle generation
- No outgassing
- High strength-to-weight ratio
- Durability in extreme environments
- Electrical conductivity.

If used properly, metal belts generate virtually no particles. A high strength-to-weight ratio minimizes belt elasticity, which could introduce a spring constant and undesired arm vibration. Metal belts provide smooth dynamics and repeatable precision control. Durability in extreme environments makes metal belts well suited for ultra-high vacuum and aggressive, corrosive process environments. Belt conductivity helps to prevent electrostatic discharge (ESD), a highly desirable property as discussed in Section 3.2. Various metal alloys with different properties are available for different environments. Metal belts do not need lubrication, which avoids contamination from outgassing lubricants. Disadvantages of metal belts are their relatively high cost and the risk of belt 'snapping' (due to near-zero elasticity) if the robot arm hits an obstacle.

Two belt types are available: endless metal belts and end-attached drive tapes. Endless metal belts are manufactured by welding the two ends of a metal tape. Endless belts can be driven either by a flat-faced friction drive pulley or a pulley with timing elements. Neither method is suitable for vacuum robots for two reasons:

- Particle generation from wear and friction between belt and friction drive pulley
- Reduced positional repeatability caused by belt creep between friction drive pulley and belt (the pulley rotates slightly faster than the belt).

Timing pulleys have teeth or pockets around the outside diameter, which engage timing holes or drive lugs on the metal belt. However, these elements are for timing purposes only, while friction is still used to drive the belt, so the risk of particle contamination persists. For these reasons end-attached drive tapes are the common solution for robot arms in vacuum. The metal belts in SCARA-type robot arms typically travel around two pulleys: one drive pulley and one driven (idler) pulley. A belt tensioner per belt is also common. The drive pulley is directly or indirectly attached to the motor through the dynamic vacuum barrier. The other pulley is driven by the belt. Pulleys are designed to match the characteristics of the metal belt and can be manufactured from a wide range of materials. Aluminum and stainless steel are common in vacuum.

End-attached drive tapes are screwed to both the drive and the idle pulley. This provides zero backlash and slippage-free motion transmission. However, the end-attachment limits the motion range to about half a rotation of the drive and idle pulleys. For the SCARA-type robot arms discussed here this is sufficient. For the remainder of the book the general term 'metal belt' is used for end-attached metal drive tapes.

Belt reliability

The following practices will maximize reliability and product life of metal belts:

- Minimize the number of pulleys
- Maximize pulley diameter
- Maximize length-to-width ratios
- Avoid reverse bending of the belt.

The number of pulleys should be minimized because each pulley bends the belt and imposes stress by unequal stretching at the inner and outer belt surfaces. Large pulley diameters mitigate that stress somewhat, while reverse bending (bending in different directions) increases that stress. A large length-to-width ratio, and a large pulley diameter combined with a small belt thickness, improve belt reliability.

Stress analysis

An estimate of the forces that would act on a belt is recommended before specifying a metal belt for a given application. A typical analysis includes the forces during standard operating conditions, worst-case scenarios (the largest expected force), the bending stress, and the total stress acting on the belt. See, for example, (Belt Technologies, 1999). Figure 4.13 illustrates the two-pulley arrangement used for each link of a SCARA-type arm.

An analysis of the bending stress and the maximum force is outlined below. First an endless belt with friction pulley is analyzed. Then the stress acting on an end-attached metal drive tape is analyzed and compared to that acting on the endless belt.

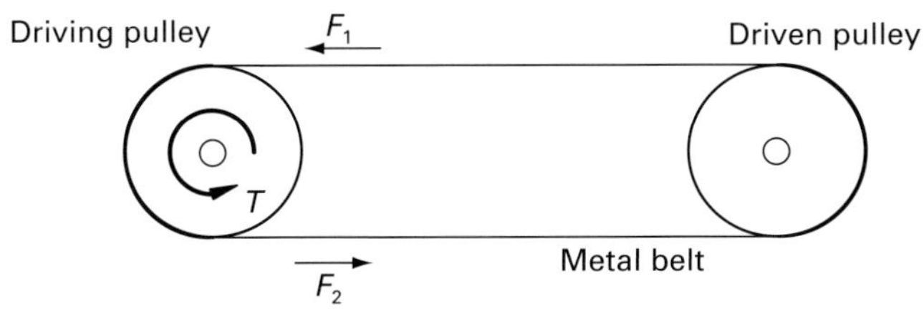

Figure 4.13 Torque and forces acting on metal belt and pulleys.

The stress force F_S acting on an endless metal belt with friction pulleys is the difference between the driving force F_1, resulting from the motor torque T, and the idle force F_2 (Figure 4.13):

$$F_S = F_1 - F_2 = \frac{T_1}{r_1} - \frac{T_2}{r_2}. \tag{4.7}$$

T_1 and r_1 are the driving torque (N·m) and radius (m) of the drive pulley, respectively. Similarly, T_2 and r_2 are the torque and the radius for the driven pulley. The maximum force acting on the belt is applied by the motor, so

$$F_{\text{max}} = F_1. \tag{4.8}$$

F_{max} can be calculated using the approximate relationship

$$\frac{F_1 - F_c}{F_2 - F_c} \approx \frac{F_1}{F_2} = e^{\mu\theta}. \tag{4.9}$$

The centrifugal force F_c acting on the belt is produced by pulley rotation and can be neglected because of the small belt mass and low speeds. The unitless friction coefficient μ is usually in the interval $0.25 \leq \mu \leq 0.45$ for friction pulleys. The 'wrapping angle' θ (rad) is determined by the contact surface between pulley and belt. With F_S from above and $F_2 = F_1 - F_S$ we get

$$F_1 = F_2 e^{\mu\theta} = (F_1 - F_S)e^{\mu\theta} \tag{4.10}$$

$$F_{\text{max}} = F_1 = F_S \frac{e^{\mu\theta}}{e^{\mu\theta} - 1}. \tag{4.11}$$

The bending stress S (N·m^2) at the pulley is given by the equation (Marcus, 2005):

$$S = \frac{E \cdot x}{(1 - u^2) \cdot r_{\text{small}}}. \tag{4.12}$$

E is Young's modulus in N·m^2, u is Poisson's ratio, and r_{small} (m) is the radius of the smallest pulley in the assembly. The belt thickness x (m) depends on r_{small} and can be obtained from the manufacturer.

For an end-attached drive tape with both ends firmly attached to the driving pulley, both pulleys rotate at the same speed, so

$$F_{\text{max}} = F_S = F_1 = F_2. \tag{4.13}$$

Equation (4.12) for the bending stress S also applies to end-attached drive tapes.

Tracking of metal belts

The tracking of a metal belt, that is, its accurate alignment with the drive and idle pulley, can be a delicate task. Metal belts practically do not stretch and therefore cannot compensate for unintended forces resulting from inaccurate tracking. Such forces include those created by pulley shaft deflection, lack of system alignment, and belt camber. Belt camber ('edge bow') is the edge deviation from a straight line, typically less than 1 mm per 1 m of

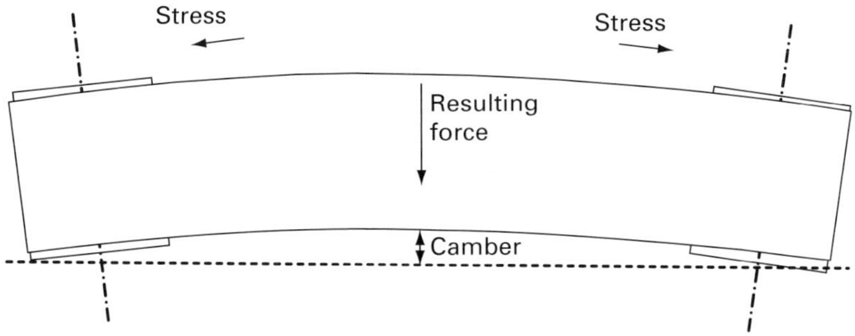

Figure 4.14 Camber (edge bow) of a metal belt.

belt length. If a belt is tensioned with a bowed edge, it will move away from the tighter edge towards the looser edge, as illustrated in Figure 4.14. The recommended method for controlling camber and similar effects is precision alignment of the pulleys. For robot arms with end-attached drive tapes stiff pulley posts facilitate belt tracking. Often the drive pulley is fixed to the motor, in which case only the idler pulley is adjusted. A properly adjusted belt tension is also important for maximizing belt reliability and life. The tension should be adjusted as low as possible without compromising robot performance.

Materials

Table 4.4 lists examples of alloys used for metal belts. The recommended alloy depends on the application. Not all alloys listed are suitable for aggressive environments.

4.5.2 Transmission with gearheads

Gearheads offer a viable transmission technology for both atmospheric robots and, with appropriate design parameters and specifications, for vacuum robots. Gearheads offer the following advantages:

- Transmission of high torques in a compact size
- Cost-effectiveness compared to other transmission methods for vacuum
- Cleanliness compatibility if properly enclosed and sealed.

The potential risks of using gearheads in vacuum are:

- Virtual leaks from air trapped in sealed gearhead housings
- Outgassing of lubrication in high vacuum
- Reduced positional accuracy and repeatability due to the backlash of most gearheads.

Solutions are available for mitigating these risks, and gearheads can be manufactured with the desired vacuum compatibility. Most virtual leaks can be prevented by selecting suitable materials and by applying the design guidelines presented in Section 4.6.2. Vacuum-compatible gearheads use vacuum-rated seals and lubricants to prevent outgassing. Ordinary lubricants begin to break down at pressures below 10^{-2} Pa ($7.5 \cdot 10^{-5}$ torr). At ultra-high vacuum pressures below 10^{-7} Pa ($7.5 \cdot 10^{-10}$ torr) special lubricants

Table 4.4. Alloys commonly used for metal belts (Source: Belt Technologies, Inc.).

Alloy	Yield strength, 0.2% off-set method ($N \cdot mm^{-2}$)	Tensile strength ($N \cdot mm^{-2}$)	Elonga-tion in 2 inches (mm)	Hardness	Tensile modulus of elasti-city ($N \cdot mm^{-2}$)	Poisson ratio	Density ($g \cdot cm^{-3}$)	Thermal expansion coefficient for 0–100 °C (K^{-1})	Corrosion resistance (low, medium, high)
301 Full Hard	1 100	1 240	5–15	RC40–45	$1.93 \cdot 10^5$	0.285	7.9	$16.9 \cdot 10^{-6}$	M
301 High Yield	1 790	1 930	1	N/A	$1.79 \cdot 10^5$	0.285	7.9	$16.9 \cdot 10^{-6}$	M
302 Full Hard	1 100	1 240	1–5	RC40–45	$1.93 \cdot 10^5$	0.285	7.9	$17.3 \cdot 10^{-6}$	M-H
304 Full Hard	1 100	1 240	1–5	RC40–45	$1.93 \cdot 10^5$	0.285	7.9	$17.3 \cdot 10^{-6}$	M-H
316 Full Hard	1 200	1 310	1–2	RC35–45	$1.93 \cdot 10^5$	0.285	7.9	$16.0 \cdot 10^{-6}$	H
716 Full Hard	1 450	1 790	5–10	RC52	$2.20 \cdot 10^5$	0.285	7.9	$10.6 \cdot 10^{-6}$	L-M
17–7 Condition C	1 275	1 480	5	RC43	$1.93 \cdot 10^5$	0.305	7.8	$15.3 \cdot 10^{-6}$	M-H
17–7 CH 900	1 655	1 720	2	RC49	$2.00 \cdot 10^5$	0.305	7.8	$10.9 \cdot 10^{-6}$	M-H
Carbon Steel SAE 1095	1 650	1 790	7–10	RC50–55	$2.07 \cdot 10^5$	0.287	7.9	$10.5 \cdot 10^{-6}$	L
Titanium 15V-3CR-3AI-3SN	1 030	1 140	11	RC35	$1.03 \cdot 10^5$	0.300	4.7	$9.7 \cdot 10^{-6}$	H

are needed. Examples are listed in Section 4.5.5. Anti-backlash gearheads are available and have been successfully used for vacuum robots (J. Thompson, personal communication, 2009). Typical contamination sources are residual lubricants, machine coolants, cleaning solvents, and fingerprints. The robot design should allow easy access to all gearheads for maintenance.

4.5.3 Bearings

Bearings are moving parts and therefore generate particles. Design objectives are therefore to minimize the number of bearings and to seal the bearings used in cleanroom robots. The sealing of bearings substantially reduces contamination. However, virtual leaks must be avoided, that is, pockets of air trapped during pumpdown in a sealed bearing. This is accomplished with release openings. Furthermore, vacuum-compliant, low-outgassing lubricants for the specified temperature range must be used. See Table 4.5 for a list of vacuum-compatible lubricants. Bearings should be thoroughly cleaned according to UHV cleaning procedures before lubrication is applied.

Bearings designed for ultra-high vacuum and harsh environments are commercially available in standard sizes and configurations (radial, angular, and four-point contact). The properties of these products includc corrosion resistance, high-temperature

Table 4.5. Vacuum-compatible lubricants and manufacturer specifications.

Product	Material or base material	Vapor pressure	Comment
TorrLube®*	Perfluoroalkylether (synthetic oil)	$1.3 \cdot 10^{-7}$ Pa (10^{-9} torr)	Wet, non-toxic, non-flammable
Fomblin® Y	Perfluoroalkylpolyether, PFPE (fluorinated polymer)	$8.0 \cdot 10^{-6}$ Pa ($6 \cdot 10^{-8}$ torr)	Wet, non-toxic, non-flammable. Tested at 25 °C
Klüberalfa® HX 83–302	Fluorinated polyether, PTFE thickener		Wet, non-toxic, non-flammable
Krytox® LVP	Perfluoroalkylpolyether (PFPE), fluorocarbon thickener	$1.3 \cdot 10^{-11}$ Pa (10^{-13} torr)	Wet, non-flammable
Krytox® 143	Fluorinated synthetic oil	$1.3 \cdot 10^{-7}$ Pa (10^{-9} torr)	Wet, non-reactive, non-flammable. Tested at 38 °C
Apiezon® Type L	Hydrocarbon	$1.0 \cdot 10^{-8}$ Pa ($8 \cdot 10^{-11}$ torr)	Wet. Tested at 20 °C
Santovac® 5	Polyphenyl ether, PPE (synthetic)	$5.3 \cdot 10^{-8}$ Pa ($4 \cdot 10^{-10}$ torr)	Wet, non-toxic. Tested at 25 °C
Christo-Lube® MCG 133	Perfluoropolyethers	$6.7 \cdot 10^{-11}$ Pa ($5 \cdot 10^{-13}$ torr)	Wet, chemically inert, non-reactive. Tested at 20 °C

* ASTM E-595 outgassing test for spacecraft use: 24 hours, 125 °C, $5 \cdot 10^{-5}$ torr; total mass loss (TML): 0.088% (required maximum 1.0%); collected volatile condensable materials (CVCM): 0.048% (required maximum 0.1%).

performance, and compatibility with aggressive chemical environments. Operation with only marginal lubrication is possible. The surface degradation and particle generation that is common in such environments is reduced with the selection of suitable materials for the bearing components. Examples are:

- Races: AISI 440C stainless steel, 17–4PH steel
- Balls: 440C stainless steel, ceramic (silicon nitride), borosilicate glass
- Separators: non-metallic composite (PEEK, Vespel®)
- Cage: PTFE or Vespel® toroid ball spacers, 300-series steel rings, stainless steel or non-metallic composite rings.

Ceramic rolling elements on hardened steel races are best with no lubrication or only marginal lubrication.

4.5.4 Electrical components

Electrical parts and components should not be used in high and ultra-high vacuum environments. Electrical connections, wire insulation materials, and solder connections can outgas in vacuum and pose a contamination risk (Aalund, 1998). See Section 4.4.7 for more details.

4.5.5 Lubricants for vacuum

Lubricants must be carefully selected for a given manufacturing process, preferably with support from the manufacturer. A limited number of suitable lubricants are available for use in vacuum environments. Most vacuum-compatible lubricants belong to one of two classes: petroleum-based lubricants or synthetic lubricants. Synthetic lubricants are designed with desired properties for certain conditions and in extreme environments generally outperform petroleum-based greases with respect to contamination and outgassing, and the required robot maintenance. Hydrocarbon-based lubricants can be too volatile, while graphite lubricants are prone to particle generation. Silicone-based lubricants can form insulating deposits that support electrostatic charge. Some grades of perfluoropolyether oils have acceptably low vapor pressures. Gold and silver plating have been used successfully and are stable at high temperatures. See also the Sematech standard 92051107A-STD and (Mattox, 1998). Table 4.5 lists examples of available vacuum lubricants.

Dry powder lubricants are not recommended for vacuum robots. However, solid lubricants (dry coatings) with low static friction may be acceptable, for example Teflon®. Vacuum robots in a central vacuum cluster tool are not directly exposed to the manufacturing process. Nonetheless, if elevated temperatures are a concern, the typical thermal limits of vacuum-compatible lubricants must be considered:

- Petroleum-based lubricants: 150 °C
- Most synthetic lubricants: 190 °C
- PPE-based lubricants: 300 °C.

PPE lubricants are a good candidate for ultra-high-vacuum environments and elevated temperature ranges. They are based on polyphenyl ether (PPE) and characterized by their molecules' high resonance energy. PPE requires significant energy to decompose, or oxidize, a molecule. The combination of temperature limit, low vapor pressure, and resistance to several reactive chemicals is desirable in vacuum and harsh environments. It is reported that electrostatic shock oxidized several tested petroleum-based and synthetic greases, creating black flakes of oxidized lubricant that contaminated a Class 10000 cleanroom. A PPE lubricant reportedly solved the problem and extended the time between maintenance (cleaning and lubrication) from days to months (Urban and Hamid, 2005).

4.6 External and internal leaks

A vacuum system can have three types of internal gas loads that impact the achievable base pressure (Marquardt, 1999): (a) residual gas in the system, (b) vapor in equilibrium with the materials exposed to vacuum, and (c) the gases introduced by virtual leaks, outgassing, and permeation. The base pressure of high-vacuum systems is mostly affected by virtual leaks (item c). A vacuum system can also have external leaks, which are minuscule, low-conductivity paths from the exterior atmospheric environment into the vacuum chamber. Both the static and the dynamic vacuum barriers of vacuum robots can have external leaks. A virtual leak, or internal leak, is a seemingly external leak that actually resides inside a vacuum enclosure. It is caused by a pocket of trapped gas that slowly evacuates into vacuum during and after pumpdown. The low-conductivity path from the gas load into the enclosure prevents a full evacuation during pumpdown. At a sufficiently low pressure the virtual leak becomes a small but continuous gas load flowing into the chamber. This can cause two common problems (Westerberg, 1999):

- External and virtual leaks prevent achievement of the desired base vacuum pressure, the minimum required pressure for a given process;
- Virtual leaks of a non-process gas can change the process conditions.

Both scenarios can impact the vacuum system's performance and compromise the manufacturing process. If the target pressure in a vacuum system cannot be achieved despite the appropriate pumping system, and if a leak test does not identify an external leak, then one or more virtual leaks are present in the vacuum chamber. The design challenge for vacuum robots is to identify and eliminate all possible sources of virtual leaks.

4.6.1 External leaks

Examples of external leaks are leaks in welding joints, scratches in O-rings, and scratches in mating surfaces and metal gaskets, but also leaks caused by contaminating particles on mating surfaces of a seal, a human hair across an O-ring, or residues from a manufacturing process on a metal gasket. Both the static and the dynamic vacuum barrier of vacuum robots are possible sources of external leaks. Not only good engineering design is needed to prevent external leaks, but also suitable procedures and processes for manufacturing, assembly, shipping, and handling. The following examples illustrate this:

- Scratches in O-rings, mating surfaces, and metal gaskets can be avoided with appropriate handling and assembly procedures and guidelines.
- Contaminating particles and machine shop residues on mating surfaces of a seal can be removed using appropriate cleaning procedures and guidelines.
- Human hair in a seal can be avoided with appropriate gowning and cleanroom guidelines.
- Welding joints can be avoided with a different design and manufacturing technology.

External leaks can be detected with vacuum (or leak) integrity tests. Helium mass spectroscopy is the prevalent method. The Sematech standard 92051107A-STD states that vacuum systems should be tested using a helium mass spectrometer leak detector with a minimum sensitivity of 10^{-11} $Pa{\cdot}m^3{\cdot}s^{-1}$ (approximately 10^{-10} $atm{\cdot}cm^3{\cdot}s^{-1}$) or better, at a pressure differential of 1 atm. When using 100% electronic grade helium and a probe speed of 2.54 $cm{\cdot}s^{-1}$ (1 $in{\cdot}s^{-1}$) or less, the system does pass the test if no leaks are detected. Leak tests can also include pressure-decay testing and seal-strength testing using burst, creep, and creep-to-failure techniques. The defect size limit can be estimated using the virtual leak diameter (VLD) method, which relates vacuum integrity to defect size using the Hagen–Poiseuille equation (Sparrow, 2004). See also Section 4.6.2.

4.6.2 Virtual (internal) leaks

This section discusses possible sources of virtual leaks (internal leaks), as well as methods to prevent them.

Sources of virtual leaks

A virtual leak is a gas source inside a vacuum chamber, typically a pocket of trapped air. During and after a pumpdown, when the chamber pressure is lower than that of the trapped air, the gas load slowly leaks into the vacuum environment. Virtual leaks can impact the vacuum of a manufacturing process in two ways, depending on the size of the gas load (Danielson, 1998; Westerberg, 1999). First, larger gas loads that are comparable to the pumping speed near the system's base pressure prevent achievement of that base pressure. This is a common problem with vacuum systems. Second, small gas loads of a non-process gas can alter the process conditions. For example, even minuscule amounts of water vapor, less than 0.01%, can change the plasma in a vacuum chamber. Furthermore, the air released from small virtual leaks can create insulating oxidation and nitride layers on metal sputtering cathodes and therefore contribute to electrostatic discharge (ESD) problems.

Typical sources of virtual leaks include:

- Blind tapped holes with non-vented screws
- Internal welding cracks

- Trapped air between concentric O-rings between two flanges
- Small cavities at the knife-edges of metal gaskets
- Rolled rather than forged flange material can lead to virtual leaks due to silicate enclosures.

Note that, although outgassing materials can have the same effect on the vacuum pressure, we distinguish outgassing from virtual leaks. Outgassing is addressed below.

Prevention of virtual leaks

It is difficult to completely prevent virtual leaks because even seemingly trivial causes like tiny gaps between flanges can establish such leaks. However, good engineering design practices and assembly procedures minimize the risk of virtual leaks. For example, engineering design reviews during the robot development phase should include an evaluation of the most common risk areas:

- Gaps
- Cracks
- Surface and seal contacts
- Trapped air pockets.

Many virtual leaks can be avoided with proper assembly and welding procedures and other bonding methods. Metal gaskets that comply with the ISO-KF standard provide axial relief holes in the center ring to evacuate gas trapped between center ring and seal (Danielson, 2004). Blind tapped holes can be avoided in several ways (Westerberg, 1999): (1) vented screws; (2) threaded holes that fully penetrate the component; (3) an escape path provided by a side vent from the threaded hole (see Figure 4.15). Welding cracks can be avoided by using monolithic work pieces, which is common practice for vacuum robots. Many vacuum chambers are manufactured from one monolithic block of aluminum. The practice of using sealed bearings, common with atmospheric robots to avoid particle contamination, is not necessarily recommended for vacuum robots. Depending on the seal type and the process conditions, air can be trapped within a seal, which would create a virtual leak.

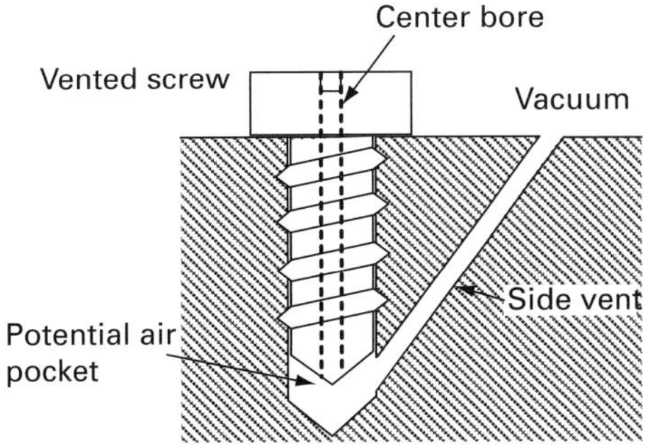

Figure 4.15 Preventing a virtual leak with a vented screw or a relief hole (side vent).

Once a vacuum system, including the robot, has been commissioned into operation, newly discovered virtual leaks may be difficult to repair. It is important to conduct helium tests during the prototype design stages to identify virtual leaks from screws or other removable parts and components. Redesigns after the product's release are expensive and generally to be discouraged. Another possibility for dealing with persistent virtual leaks is to first pump down the chamber and then fill it with argon or other inert gas that is compatible with the manufacturing process. The inert gas will reduce the vacuum pressure and therefore prevent virtual leaks from releasing gas into the chamber. Over time this is an expensive compromise.

Detecting virtual leaks

Virtual leaks are difficult to detect and locate, and are sometimes 'invisible' even to helium tests. If direct measurements are not possible, secondary indicators are needed, for example abnormal pressure curves during pumpdown, or residual pressures above the specified base pressure (Helgeland, 2002). Some virtual leaks produce a discrete series of gas bursts, similar to a bubble series. Analog pressure gauges can measure such bursts. For example, if an otherwise constant pressure of $2.6 \cdot 10^{-3}$ Pa ($2.0 \cdot 10^{-5}$ torr) shows a series of spikes of $3.1 \cdot 10^{-3}$ Pa ($2.3 \cdot 10^{-5}$ torr), then a bursting virtual leak is the likely cause (Figure 4.16). The spike frequency often increases or decreases with the ambient temperature (Danielson, 2003). Such observations can help to determine the presence of virtual leaks.

Virtual leak diameter

An estimate of the virtual leak size can be helpful in identifying a leak. The virtual leak diameter (VLD) technique was developed in the early 1990s at Ford Industries and Uson L.P. The VLD technique uses the measured mass flow from a suspected virtual leak and

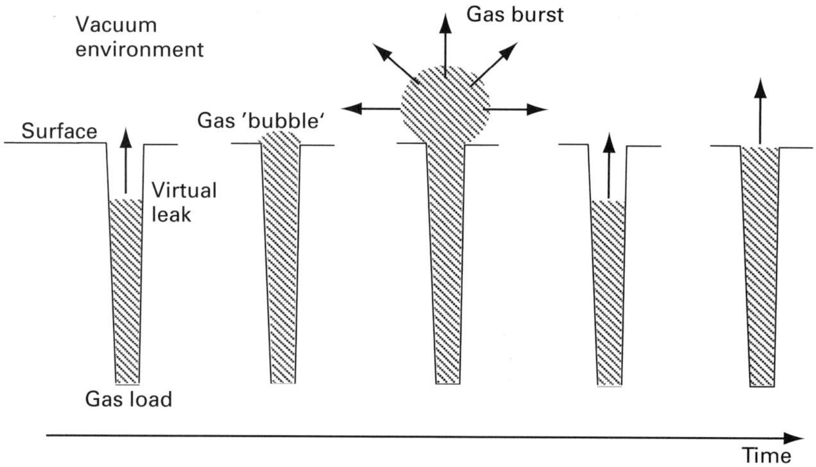

Figure 4.16 A gas burst from a virtual leak over time, after Danielson (1998).

assumes that the leak path obeys the Hagen–Poiseuille flow equation. This equation describes the laminar and constant volume flow of an incompressible, uniform, and viscous liquid through a small cylindrical tube. The tube is assumed to have a circular and constant cross section. The technique can be applied, using accurate gauges, to detect the presence of virtual leaks and estimate their diameters.

The Hagen–Poiseuille equation describes the stationary, laminar volume flow Φ (Sutera and Skalak, 1993):

$$\Phi = \frac{dV}{dt} = \frac{dx}{dt}A = \frac{dx}{dt}\pi r^2 = v\pi r^2 = \frac{r^2\Delta p}{8\eta l}\cdot\pi R^2 = \frac{\pi r^4(p_1 - p_0)}{8\eta l} = \frac{2\pi d^4(p_1 - p_0)}{\eta l}. \tag{4.14}$$

The volumetric flow dV/dt ($m^3 \cdot s^{-1}$) is the volume of a gas or fluid that passes through a given volume per time interval t (s). The parameter d (m) is the virtual leak diameter, v ($m \cdot s^{-1}$) is the median fluid velocity along the length of the tube, x (m) is the variable length of the tube, r (m) is the internal radius of the tube, $\Delta p = p_1 - p_0$ in (Pa) is the difference between the absolute internal pressure p_0 and absolute external pressure p_1 along the tube, η (Pa·s) is the dynamic viscosity, and l (m) is the length of the virtual leak path, or the thickness of the vacuum barrier.

We are interested in estimating the virtual leak diameter d. If a measurement of the stationary volume flow V is not available, a secondary indicator, the mass flow rate dM/dt, can be computed instead, using numerical differentiation of the measured mass flow M ($kg \cdot s^{-1}$) of the virtual leak over time t:

$$\frac{dM}{dt} = \frac{dV}{dt}\rho = \frac{2\pi d^4 \Delta p}{\eta l}\rho. \tag{4.15}$$

The virtual leak diameter d can then be determined by solving the Hagen–Poiseuille equation for d:

$$d = \sqrt[4]{\frac{dM}{dt}\cdot\frac{\eta l}{2\pi\rho\Delta p}}. \tag{4.16}$$

The dynamic viscosity η is known for relevant gases. For example, at 20 °C the dynamic viscosity of air is $\eta_{air} = 1.8 \cdot 10^{-5}$ Pa·s, and for helium it is $\eta_{He} = 1.9 \cdot 10^{-5}$ Pa·s. The mass flow rate dM/dt can be numerically derived from the measured mass flow M. The pressures p_0 and p_1 can be measured. The density ρ is known for all relevant gases, for example $\rho_{air} = 1.16\ kg \cdot m^{-3}$ and $\rho_{He} = 0.17\ kg \cdot m^{-3}$.

The VLD method was first used to assess extremely small leaks with molecular regime flow rates in engine fuel systems. The research question was whether the extremely small VLDs of virtual leaks in fuel systems were sufficiently large to be fluid leaks, or if only a gas could leak into the vacuum enclosure. The estimated VLD was reported as $2.0 \cdot 10^{-5}$ m, and the estimated leak rate as a minuscule $1.0 \cdot 10^{-6}\ cm^3 \cdot s^{-1}$ atm (A. Campbell, personal communication, 2007).

Example 4.3: impact of a virtual leak on a chamber's base pressure

The impact of gas loads from a virtual leak on the base pressure of a vacuum system is determined using the law of ideal gases. Consider a non-vented screw in a blind tapped hole inside a vacuum chamber. The desired base pressure is $p_2^* = 1.33 \cdot 10^{-6}$ Pa ($1.0 \cdot 10^{-8}$ torr). The screw traps a pocket of air at atmospheric pressure at the end of the tapped hole (compare Figure 4.15). The trapped air is in 'State 1.' The pocket becomes a virtual leak that slowly releases its entire gas load into the vacuum chamber. After its release this amount of air is in 'State 2.' What is the resulting new pressure in the chamber?

Notation:

p_1, V_1, T_1 ... pressure, volume, and temperature in State 1
p_2, V_2, T_2 ... pressure, volume, and temperature in State 2

Specified parameters:

$d = 5.0$ mm ... tapped hole's diameter
$h = 2.0$ mm ... the height of the pocket left by the inserted screw
$p_1 = 101\,325$ Pa ... atmospheric pressure of the trapped air (State 1)
$V_2 = 0.1$ m^3 ... vacuum chamber volume

The volume of the air pocket is

$$V_1 = \pi \cdot \left(\frac{d}{2}\right)^2 \cdot h = \pi \cdot \left(\frac{5}{2}\right)^2 \text{m}^2 \cdot 0.002 \text{ m} = 3.927 \cdot 10^{-8} \text{ m}^3. \tag{4.17}$$

We use the thermodynamic equation for ideal gases,

$$p_1 \cdot \frac{V_1}{T_1} = p_2 \cdot \frac{V_2}{T_2}. \tag{4.18}$$

Assuming a constant temperature $T_1 = T_2$, the new pressure in the chamber is

$$\begin{aligned} p_2 = p_1 \cdot \frac{V_1}{V_2} &= 101\,325 \text{ Pa} \cdot \frac{3.927 \cdot 10^{-8} \text{ m}^3}{0.1 \text{ m}^3} \\ &= 3.979 \cdot 10^{-2} \text{ Pa} = 2.985 \cdot 10^{-4} \text{ torr}. \end{aligned} \tag{4.19}$$

This is orders of magnitude outside the specified base pressure p_2^* and creates a process condition under which the manufacturing process may malfunction. This demonstrates that the prevention of virtual leaks is critical and must be addressed during the engineering design process. □

4.7 Materials and surface finishes

Materials and surface finishes of vacuum robots impact the performance of vacuum systems in several ways, including a reduction in the achievable vacuum pressure and the contamination of substrate surfaces through condensation (Strong, 1938). The engineering challenges related to materials and surface finishes include:

- Selection of suitable materials and surface finishes, based on performance requirements, cost, and appearance
- Quantification of a robot's impact on a tool's overall cleanliness performance.

Common outgassing sources are materials with high vapor pressure, for example non-metallic seals; moisture in lubricants, and adhesives; plastics, and some metals at sufficiently high vacuum pressures (Marquardt, 1999). A method for estimating the outgassing of a given material is proposed below. The method is based on the ideal gas law and utilizes a material database established for the NASA space program. A second concern in high and ultra-high vacuum is the permeation of gas loads from the ambient atmosphere through a vacuum barrier into a vacuum chamber. No material provides a perfect vacuum barrier. Particular attention should be paid to the permeation of elastomer and other non-metallic seals.

4.7.1 Vapor pressure

All solid and liquid materials tend to evaporate to a gaseous form, and all gases have a tendency to condense back to a solid or liquid state. This behavior is influenced by the material's vapor pressure, an important material property for vacuum systems. For a given temperature all solid and liquid materials have a specific vapor pressure at which material will vaporize as many molecules as it absorbs. The following definition applies.

Definition 4.3: The vapor pressure of a material is the pressure, at a given temperature at which the material's evaporation and condensation are in a dynamic equilibrium.

High or ultra-high vacuum pressures can approach or exceed, towards zero absolute pressure, the vapor pressures of even solid materials. Vapor pressures and outgassing increase with increasing temperature. The boiling point of a liquid is the temperature at which the vapor pressure equals the ambient pressure. The vapor pressure of water is shown in Figure 4.17 as a function of water temperature. Note that the vapor pressure at 100 °C is 101 325 Pa (760 torr), which of course is the boiling point of water at sea level.

4.7.2 Outgassing

Several definitions of outgassing are available, for example the "slow release of a gas that was trapped, frozen, absorbed or adsorbed in some material" (Strong, 1938). The SEMI standard SEMI F51 defines it more specifically as a "process whereby molecules of air or other gases adhere to the surface of the vacuum vessel or component therein and become liberated under vacuum conditions." In the present context of cleanroom robots, outgassing is described as gas desorption from the surface as well as from the bulk of the

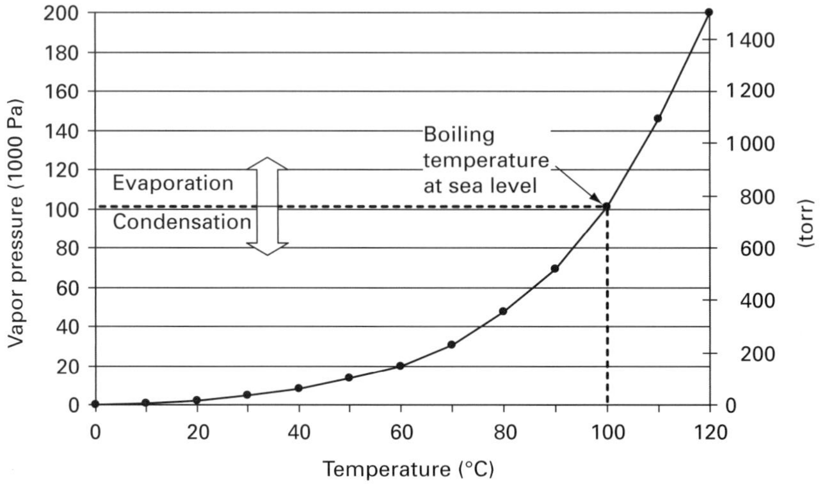

Figure 4.17 Vapor pressure of water as a function of temperature.

material (from where gas diffuses to the surface, adsorbs, and then desorbs). Outgassing is different from a virtual leak in that it does involve a state transition.

Definition 4.4: Outgassing is the release of gas from a surface under vacuum pressure. The outgassing rate is the amount of gas released in a given time period.

The outgassing rate O_R is a quantitative measure of outgassing:

$$O_R = \frac{p \cdot V}{A \cdot t}, \tag{4.20}$$

where V is the gas volume that outgasses during elapsed time t from a surface area A at pressure p. A popular non-SI unit for outgassing rate is (torr · liter · cm^{-2} · s^{-1}). The SI-based unit is $Pa{\cdot}m^3{\cdot}m^{-2}{\cdot}s^{-1} = Pa{\cdot}m{\cdot}s^{-1}$.

$$\frac{\text{torr} \cdot \text{liter}}{\text{cm}^2 \cdot \text{s}} = 1333.3 \frac{\text{Pa} \cdot \text{m}^3}{\text{m}^2 \cdot \text{s}}. \tag{4.21}$$

Outgassing can impact the performance of a vacuum system in several ways, in particular:

- Reduction of the achievable vacuum pressure
- Contamination of surfaces through condensation.

The outgassing rate increases with the ambient temperature, as well as with the vapor pressure of the gas. For solid materials the manufacturing method and preparation can substantially reduce outgassing. This is discussed below.

Figure 4.18 shows a qualitative pumpdown curve for high-vacuum systems. The actual time required and the final pressure achieved depend on the vacuum chamber volume, materials, and pumps. Note the potentially extremely long time periods. First the gas volume is removed from the chamber (exponential curve). At low and medium

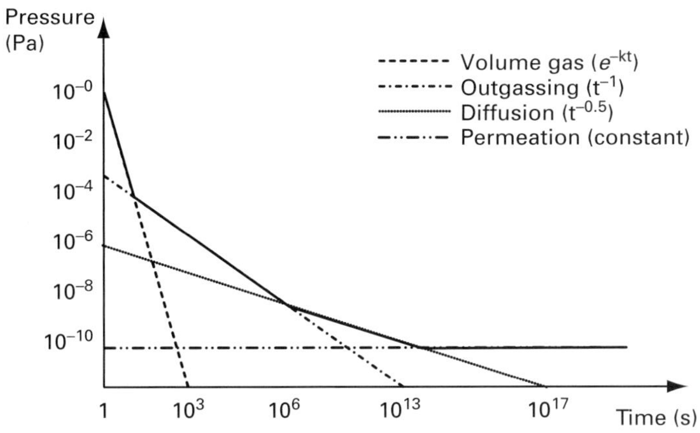

Figure 4.18 Qualitative pumpdown curve.

vacuum pressure the outgassing of 'surface gas,' or surface adsorption, begins (proportional to 1/time). At high vacuum pressure diffused gas begins to outgas (proportional to $1/\sqrt{\text{time}}$). Gas permeation takes effect at an ultra-high pressure and contributes a constant gas load to the chamber. The effective pumpdown curve is a combination of these effects (R. Bergner, personal communication, 2008; Danielson, 1998).

Other processes that are related to outgassing are:

- Adsorption, the accumulation of a gas or liquid on the surface of a solid material
- Absorption, the diffusion of a gas into a liquid, or of a liquid into a solid; desorption is the reverse of this process
- Sublimation, the transition from a solid to a gaseous state without an intermediate liquid state
- Evaporation, the transition from a liquid state to a gaseous state without reaching the boiling temperature (the opposite of condensation)
- Desorption, the transition from an adsorbed or absorbed state on a surface to a gaseous or liquid state
- Slow chemical reactions of solid or fluid materials into a gaseous state.

Boiling is not considered to be outgassing because it occurs much more rapidly.

Low-outgassing materials

Experimental, ultra-low outgassing rates have been reported for three specially treated materials (Watanabe, 2001): pure copper (oxygen-free, high-conductivity), a chromium–copper alloy with 0.6% chromium and 99.4% copper, and stainless steel 304. The materials were first baked at temperatures of at least 250 °C (copper) or 400 °C (chromium–copper alloy, stainless steel) before the outgassing rates were measured. The copper and chromium–copper alloy reached an outgassing rate of 10^{-12} Pa·m·s^{-1} (hydrogen equivalent). The outgassing rate of the stainless steel chamber was over an order of magnitude higher. Another excellent outgassing rate of $2.67 \cdot 10^{-13}$ Pa·m·s^{-1} or $2 \cdot 10^{-15}$ torr · liter · cm^{-2} · s^{-1} for 300-series stainless steel has been reported (Sasaki, 2007). It was

Table 4.6. Approximate outgassing rates after one hour in vacuum at room temperature (Danielson, 2003; Marquardt, 1999).

Material	Outgassing rate ($Pa \cdot m \cdot s^{-1}$)	Outgassing rate ($torr \cdot liter \cdot s^{-1} \cdot cm^{-2}$)
Aluminum (fresh)	$9 \cdot 10^{-7}$	$6.75 \cdot 10^{-9}$
Aluminum (20 hours at 100 °C)	$5 \cdot 10^{-12}$	$3.75 \cdot 10^{-14}$
Stainless steel (304)	$2 \cdot 10^{-6}$	$1.50 \cdot 10^{-8}$
Stainless steel (304, electropolished)	$6 \cdot 10^{-7}$	$4.50 \cdot 10^{-9}$
Stainless steel (304, mechanically polished)	$2 \cdot 10^{-7}$	$1.50 \cdot 10^{-9}$
Stainless steel (304, electropolished, 30 hours at 250 °C)	$4 \cdot 10^{-10}$	$3.00 \cdot 10^{-12}$
Teflon®	$8 \cdot 10^{-6}$	$6.00 \cdot 10^{-8}$
Viton® A (fresh)	$2 \cdot 10^{-4}$	$1.50 \cdot 10^{-6}$
Viton® (baked)	$5 \cdot 10^{-6}$	$4.00 \cdot 10^{-8}$

accomplished in a laboratory environment through a combination of electropolishing and vacuum baking. Electropolishing created a chromium-enriched surface by removing nickel (Ni), iron (Fe), and manganese (Mn) from the surface. It also reduced the atomic surface area closer to its geometric surface area. A very smooth surface with a submicron roughness traps fewer particles and thus increases cleanliness. This can be accomplished with mechanical polishing followed by electropolishing and vacuum baking.

The above experimental outgassing results were achieved in a laboratory environment. The outgassing rates of common materials used for vacuum systems and robots are orders of magnitudes higher. Outgassing data by material is available and should be consulted in the design of vacuum systems and robots. Variations in the test data are due to the number of parameters influencing the non-trivial measurements. Nonetheless, such data allows a rough estimate of the gas load a vacuum robot might release. The gas load can then be reduced by design modifications, and perhaps by using materials with better vacuum properties. Table 4.6 lists outgassing rates of relevant materials.

NASA database of low-outgassing materials

Datasets with outgassing rates obtained using a consistent and repeatable test method are rare. One example is a database established by NASA (Campbell and Scialdone, 1997). It represents outgassing in terms of the 'total mass loss' in percent. Conversion from NASA's unitless entities to the outgassing rate is discussed below.

Outgassing is a concern to space missions, where this and condensation can obscure telescopes, thermal radiators, and solar cells, and ruin an expensive mission. Starting in the late 1960s NASA conducted a comprehensive outgassing study of over 10 000 materials and commercial products for space missions. A standard test method, ASTM standard E 595-77/84/90, was developed by the American Society for Testing and Materials (ASTM), based on an apparatus developed by the Stanford Research Institute (SRI). The test method measures total mass loss (TML) and collected volatile condensable materials (CVCM). Specified test procedures were aimed at providing consistent results. All material samples had a nominal mass of $2.5 \cdot 10^{-4}$ kg and were pre-conditioned for 24 hours at 25 °C, 50% relative humidity, and standard atmospheric pressure. The samples were then weighed and placed at atmospheric pressure in a copper compartment that was heated to 398 K (125 °C) for 24 hours. The sample was heated by conduction and radiation. All outgassed volatile material could only escape through a 6.3 mm (0.25 in) diameter exit port and was collected by a chromium-plated disc at a line-of-sight distance of 12.7 mm (0.5 in) from the exit port. The collector's temperature was 298 K (25 °C), so a significant portion of the escaping volatile material was collected if its condensation temperature was 298 K (25 °C) or above. After the 24-hour heat exposure the samples were weighed again and the mass loss of the sample was determined. The lost mass in percent is the TML. Similarly, the weight gain of the collector in percent, i.e., the mass of the condensed materials, is the CVCM. The following definition is after NASA (Campbell and Scialdone, 1997).

Definition 4.5: Low-outgassing materials have a total mass loss (TML) of 1.0% or less, and a collected volatile condensable materials (CVCM) of 0.1% or less.

The NASA database of outgassing data is available online at http://outgassing.nasa.gov. It can serve as a guideline in the design of vacuum systems. The data can be used in various ways:

- A material qualifies as low-outgassing if its TML and CVCM meet the low-outgassing criterion in Definition 4.5. The outgassing rate is not needed.
- The outgassing rate of a material listed in the database can be estimated from the TML or CVCM using the method outlined below.

Note that outgassing measurements can vary and that the computed outgassing rate is only a rough estimate.

Converting TML or CVCM to outgassing rate

TML and CVCM can be converted to the outgassing rate O_R in Eq. (4.20) using the ideal gas law. The TML is the outgassed mass m_1 that passes through an area A_1 during the time period $\Delta t = 24$ hours: TML = m_1/A_1 in ($\mathrm{kg \cdot m^{-2}}$). The CVCM is the mass m_2 collected on an area A_2: CVCM = m_2/A_2. The molar mass M ($\mathrm{kg{\cdot}mol^{-1}}$) of the outgassed material is needed. With the nominal outgassing area A_{nom} and the nominal time period t_{nom} specified by the ASTM standard E 595-77/84/90 the outgassing rate is defined as

$$O_R = \frac{m_{out}}{A_{nom}} \cdot K \qquad \left(\mathrm{torr \cdot liter \cdot cm^{-2} \cdot s^{-1} = Pa \cdot m \cdot s^{-1}}\right). \tag{4.22}$$

See Appendix A for unit conversion factors. K is a factor specific to a given material and is derived below. The nominal mass m_{nom} of the material sample before the test is 0.25 g. The ASTM standard allows 0.1 to 0.3 g. The outgassed mass m_{out} is

$$m_{\mathrm{out}} = m_{\mathrm{nom}} \cdot \mathrm{TML} = 0.25\ g \cdot \mathrm{TML}. \tag{4.23}$$

It is well known that under standard conditions (temperature $T = 273.16$ K, pressure $p_{\mathrm{std}} = 101\,325$ Pa), one mole (gram molecular weight) of a gas has a volume $V_{\mathrm{std}} = 22.4$ liter·mol^{-1}. The outgassing rate is therefore

$$\begin{aligned} O_{\mathrm{R}} &= \frac{m_{\mathrm{nom}} \cdot \mathrm{TML}}{A_{\mathrm{nom}}} \cdot K, \text{ with } K = \frac{p_{\mathrm{std}} \cdot V_{\mathrm{std}}}{M_{\mathrm{mat}} \cdot \Delta t} \\ O_{\mathrm{R}} &= \frac{2.5 \cdot 10^{-1}\mathrm{g} \cdot \mathrm{TML}}{0.3167\mathrm{cm}^2} \cdot \frac{760 \text{ torr} \cdot 22.4 \text{ liter} \cdot \mathrm{mol}^{-1}}{M(\mathrm{g} \cdot \mathrm{mol}^{-1}) \cdot 86\,400 \text{ s}} \\ &= \frac{\mathrm{TML}}{M} \cdot 0.1555 \text{ torr} \cdot \text{liter} \cdot \mathrm{cm}^{-2} \cdot \mathrm{s}^{-1}. \end{aligned} \tag{4.24}$$

Note that an outgassing rate based on measurement data should be considered a rough estimate.

Example 4.4: outgassing rate of aluminum

Two aluminum samples in the NASA database are LAMINATE SHIM ALUMINUM and PAA-CORE ALUMINUM HONEYCOMB. The molar mass of aluminum is

$$M_{\mathrm{Al}} = 2.698 \cdot 10^{-2}\,\mathrm{kg} \cdot \mathrm{mol}^{-1} = 26.98\,\mathrm{g} \cdot \mathrm{mol}^{-1}. \tag{4.25}$$

A rough estimate of the outgassing rate for each aluminum sample is now calculated.

(A) LAMINATE SHIM ALUMINUM (sample ID GSC10284): TML = 0.0005, or 0.05%

$$\begin{aligned} O_{\mathrm{R,Al}} &= \frac{\mathrm{TML}_{\mathrm{Al}}}{M_{\mathrm{Al}}} \cdot 0.1555 \\ &= \frac{0.0005}{26.98} \cdot 0.1555 \\ &= 2.88 \cdot 10^{-6} \text{torr} \cdot \text{liter} \cdot \mathrm{cm}^{-2} \cdot \mathrm{s}^{-1}. \end{aligned} \tag{4.26}$$

(B) PAA-CORE ALUMINUM HONEYCOMB (sample ID GSC18471): TML = 0.0006

$$\begin{aligned} O_{\mathrm{R,Al}} &= \frac{\mathrm{TML}_{\mathrm{Al}}}{M_{\mathrm{Al}}} \cdot 0.1555 \\ &= \frac{0.0006}{26.98} \cdot 0.1555 \\ &= 3.46 \cdot 10^{-6} \text{torr} \cdot \text{liter} \cdot \mathrm{cm}^{-2} \cdot \mathrm{s}^{-1}. \end{aligned} \tag{4.27}$$

The TML or CVCM values can vary significantly depending on mix, cure, and changes in formulation of materials over the years. Therefore this method provides only a first step towards the selection of vacuum-compatible materials. □

Example 4.5: commercially available lubricant

TorrLube® is a commercially available lubricant for vacuum applications. An outgassing test of TorrLube® was conducted according to the ASTM standard E-595 to qualify the lubricant as a 'low-outgassing' material for spacecraft use. According to the manufacturer a TML of 0.088% and CVCM of 0.048% were measured. Therefore the lubricant meets the requirements. □

4.7.3 Permeation

Permeation is the penetration of a solid material by a permeate, either gas, vapor, or fluid. Here permeation refers to ambient air entering the interior of a vacuum chamber through the vacuum barrier. The air penetrates elastomeric O-rings and other non-metallic seals and eventually reaches the vacuum environment. Permeation remains constant over time, while outgassing decreases. The impact of permeation must be evaluated during the design process by comparing the available permeation constants for candidate materials.

Permeation can be divided into three phases: (1) the permeating gas is adsorbed at the exterior surface of the vacuum barrier; (2) it diffuses through the material and penetrates the material through molecular pores; (3) it desorbs and leaves the interior surface as a gas. Figure 4.19 shows the permeation and outgassing rates of unbaked and baked Viton®, a popular elastomer. The curves were acquired during a controlled experiment and are specified with respect to unit length instead of unit area. The permeation is constant over time and is small compared to the outgassing of unbaked O-rings, but is larger than the outgassing of baked O-rings. The vacuum-baked O-ring outgasses by two to three orders of magnitude less than the untreated O-ring.

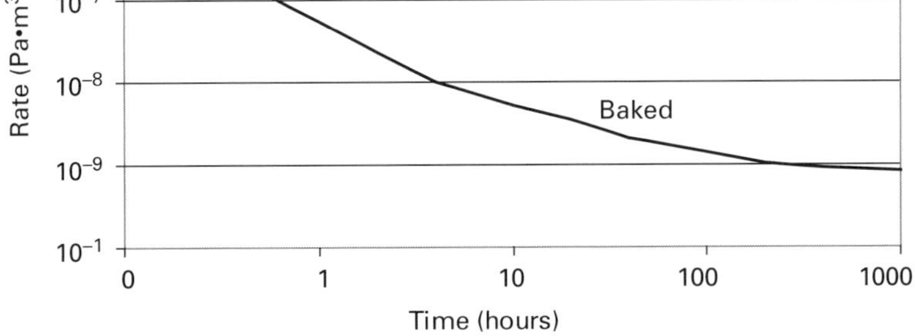

Figure 4.19 Permeation and outgassing rates of Viton® (Danielson, 2000).

Table 4.7. Permeation constants for materials (in 10^{12} $m^3 \cdot m^{-2} \cdot Pa^{-1} \cdot s^{-1} = 10^{12}$ $m \cdot Pa^{-1} \cdot s^{-1}$).

Material	He	N_2	CO_2	H_2O
Buna-N®	0.75	0.02	0.56	75.01
Viton®	0.90 – 1.73	0.02 – 0.05	0.23 – 0.60	3.90
Kalrez®	8.40	0.23	1.88	
Chemraz®	10.73	0.66		
Silicone	23.25 – 24.75	7.50 – 12.00	45.00 – 225.02	300.02 – 750.06
Teflon®	5.10	0.11 – 0.24	0.90	2.70

The permeation constants of selected materials, measured with four different permeating gases, at standard temperature and pressure (STP) are listed in Table 4.7 (Danielson, 2001b). A smaller coefficient means that the material allows less permeation for a given gas. The determination of gas permeation parameters is addressed, for example, by Alentiev and Yampolskii (2006).

4.7.4 Materials for vacuum robots

The above discussion underlines the importance of selecting suitable materials for exposure to the vacuum or process environment. Every material must be evaluated, ideally together with the end-user, with respect to any physical or chemical requirements imposed by the manufacturing process. The following summarizes desirable or required material properties that are essential for most vacuum systems. See also (Chambers et al., 1998; Danielson, 2003; Halliday, 1987a).

Desired mechanical properties:

- The materials retain the specified elastic, plastic, fluid, or other properties under process conditions.
- Materials used for the vacuum barrier withstand the forces exerted by the vacuum pressure.
- Materials for machined parts have suitable properties for machining.

Desired thermal properties:

- The vapor pressure remains low for the entire temperature range.
- Thermal expansion does not differ substantially from that of adjacent materials, in particular at joints.

Desired outgassing properties:

- Materials don't have cracks, crevices, or other minuscule cavities that will trap outgassing substances, e.g., cleaning solvents, and become outgassing sources during pumpdown.
- Surface and bulk desorption rates don't exceed the specified gas loads.
- Materials are not porous and do not absorb water.

Metals for vacuum robot parts and components

A design objective for vacuum robots is minimization of the total surface area to be exposed to vacuum. A minimal atomic surface area achieves the lowest absorption and desorption rates and can be obtained with electropolishing. This is recommended for machined and welded metal parts. Most structural parts of vacuum robots in electronics manufacturing are made of metals, in particular stainless steel and aluminum alloys (Chambers et al., 1998; Danielson, 2003; Halliday, 1987a):

- Aluminum and aluminum alloys are cost-efficient, easy to machine, and suitable for high and ultra-high vacuum. The parts exposed to vacuum are uncoated. Outgassing rates are very low if the zinc content is low. Anodizing is not recommended for vacuum because the porous oxide film absorbs water vapor and later releases it. Alodining, a chemical treatment process for corrosion protection and surface preparation, may be acceptable. Welding is a risk, but is rarely needed for vacuum robots.
- Stainless steel, an iron–carbon alloy, is excellent for high and ultra-high vacuum. It can be machined and has very low outgassing rates. The 300 series is common, for example the 304 and 316 series, and the low-carbon alloys 304L and 316L for ultra-high vacuum. The Sematech standard 92051107A-STD recommends that vacuum surfaces be mechanically polished to a 0.813 μm roughness (R_a) finish, followed by electropolishing to a 0.254 μm R_a finish. This reduces the atomic surface towards the geometric surface area. Welding is a risk, but is rarely needed for vacuum robots.
- Mild steel, with a low carbon content of up to 0.3%, is suitable for high vacuum above 10^{-4} Pa ($7.5 \cdot 10^{-7}$ torr). At lower pressures its permeability to hydrogen can be a risk.

Materials with vapor pressures above the base pressure in a vacuum chamber should be avoided. Examples are cadmium, used for plated steel screws; zinc, present in brass; sulfur and selenium in 303 stainless steel. Casting of steel is preferred over rolling, which can create virtual leaks. It is desirable that the robot surfaces exposed to vacuum have a finish similar to the interior chamber surfaces, not only to match their technical properties, but also to improve the visual appearance to the user.

Metals for vacuum seals

Metal seals and gaskets outperform non-metal O-rings in high and ultra-high vacuum. They avoid the permeation and gas loads from plastics and elastomers. Metal gaskets are commercially available. Parts from different manufacturers should not be mixed. Common metals for gaskets and seals include copper, aluminum, indium, and gold.

Copper and oxygen-free, high-conductivity (OFHC) copper can be machined and have good corrosion resistance. Baking can cause oxidation. Aluminum wire rings are relatively inexpensive and can be heated up to 200 °C. Indium wire is soft and is pressed between flat mating surfaces. It can continue to flow after the initial tightening. Gold wire is used for ultra-high-vacuum seals between two flat mating surfaces. It can be heated up to 450 °C.

Table 4.8. Permeation and outgassing rates of O-ring materials (Danielson, 2004).

Elastomer	Permeation by water ($m \cdot Pa^{-1} \cdot s^{-1}$)	Permeation by helium ($m \cdot Pa^{-1} \cdot s^{-1}$)	Outgassing ($Pa \cdot m \cdot s^{-1}$)	Temperature limit (°C)
Butyl	$6.67 \cdot 10^{-8}$	$4.00 \cdot 10^{-7}$	$1.33 \cdot 10^{-4}$	86
Nitrile	$1.33 \cdot 10^{-6}$	$1.20 \cdot 10^{-6}$	$1.33 \cdot 10^{-5}$	135
Fluoro-elastomer	$8.00 \cdot 10^{-8}$	$6.67 \cdot 10^{-7}$	$4.00 \cdot 10^{-7}$	150
Perfluoro-elastomer	$2.67 \cdot 10^{-7}$	$1.33 \cdot 10^{-5}$	$4.00 \cdot 10^{-7}$	200

Non-metal seal and gasket materials

Non-metal gaskets and O-rings are convenient and re-usable alternatives to metal gaskets for low- and medium-vacuum applications. Outgassing and permeation are risks. The materials must be carefully selected based on both application requirements and manufacturer specifications. Table 4.8 lists vacuum properties of some elastomers used for vacuum seals. The permeation and outgassing rates were measured at 23 °C and 50% air humidity. Note that the measurement unit is per O-ring length (m^{-1}), not O-ring surface area.

The following are examples of commercially available elastomers used for vacuum seals:

- Viton® is a fluoroelastomer. It is bakeable to 200 °C and suitable up to high vacuum pressures. The Sematech standard 92051107A-STD approves fluoroelastomers for DI applications.
- Kalrez® and Chemraz® are perfluoroelastomers and are often considered as alternatives to Viton®.
- Nitrile rubber (e.g., Buna-N®) demountable seals and O-rings can be used for moderate vacuum.

Plastics are not always suitable for vacuum systems, due to their relatively high gas permeability and desorption rates. However, certain properties and the cost-efficiency of plastics can be compelling. After sufficient vacuum treatment to remove gas loads, the following plastics may be suitable for moderate vacuum applications:

- Polyethylene can be suitable for medium to high vacuum pressures, to approximately 0.133 Pa (10^{-3} torr).
- Polycarbonate is used for windows, instead of glass materials such as Pyrex® or quartz, for vacuum pressures to approximately $1.33 \cdot 10^{-3}$ Pa (10^{-5} torr).
- Kel-F®, or chlorotrifluoroethylene, is a thermoplastic with low outgassing and absorption rates. It has an operating temperature range from –240 °C to 204 °C.
- Kynar®, or polyvinylidene fluoride, is a thermoplastic with an acceptable outgassing rate for some vacuum applications.
- G10 glass epoxy is acceptable if the relatively long time for initial outgassing can be tolerated.

Kel-F®, Kynar®, and G10 glass epoxy are cost-efficient alternatives to Vespel®. Teflon®, a polytetrafluoroethylene (PTFE), is a synthetic fluoropolymer that is not suitable for vacuum barriers. Although it is self-lubricating, relatively low-outgassing, and temperature-resistant, the high permeability eliminates it as a gasket material. It is common as a coating material.

Vacuum-compatible lubricants

Vacuum-compatible lubricants suitable for vacuum robotics are listed in Table 4.5. For example, Krytox®, a liquid perfluorinated polyether, is available in vacuum-compatible versions. TorrLube®, a perfluoroalkylether, is a viscous fluid, highly distilled and submicron-filtered, and is suitable for ultra-high vacuum. Polyphenyl ether (PPE) is suited for ultra-high vacuum. It exhibits good thermal and chemical stability and a low vapor pressure. Fomblin® is a perfluoropolyether that exhibits good chemical stability and resistance to many aggressive chemicals.

Vacuum-compatible epoxies and sealants

Epoxies are rarely used in high-vacuum applications, because they cure by evaporation of the solvent. The curing tends to start from the outside and progresses inward. The solvent evaporation can last for months. However, in certain situations epoxies may offer viable solutions. Examples of vacuum-compatible products include:

- Torr Seal® by Varian is a solvent-free epoxy resin for pressures down to about 10^{-7} Pa ($7.5 \cdot 10^{-10}$ torr) and temperatures from –45 °C to 120 °C.
- KL-320K by the Kurt J. Lesker Co. is suitable for pressures down to $1.33 \cdot 10^{-5}$ Pa (10^{-7} torr). KL-325K is an electrically conductive epoxy specified for pressures down to $1.33 \cdot 10^{-2}$ Pa (10^{-3} torr).

The above specifications have been provided by the manufacturers.

4.8 Assembly and installation of vacuum robots

This section summarizes basic guidelines for the assembly of vacuum robots and their preparation for installation in vacuum chambers. Assembly and installation should be performed according to strict procedures in order to protect the robot system from contamination and damage that could compromise the specified cleanliness and vacuum integrity. Even minor damage and tiny amounts of contaminants can limit the achievable pressure in a vacuum chamber. It has been shown that strict precautions and procedures are necessary to achieve a profitable electronics manufacturing process (Christensen et al., 2008).

4.8.1 Cleaning

All machined parts and subassemblies must be cleaned prior to robot assembly. Cleaning procedures for ultra-high vacuum specify how to remove residual lubricants, machine

Example 4.6: impact of a fingerprint on vacuum

The impact of a fingerprint on vacuum base pressure can be substantial. The gas load of one fingerprint is about $1.33 \cdot 10^{-6}$ Pa·m^3·s^{-1} (10^{-5} torr·liter·s^{-1}), depending on its size and moisture (Danielson, 2001a). This seemingly small leak rate is unacceptable for many manufacturing processes in high vacuum. The pumping speed necessary to compensate for such a leak rate can be determined with the basic equation for vacuum systems,

$$q_{gas} = s_{pump} \cdot p \tag{4.28}$$

where q_{gas} is the gas load (Pa·m^3·s^{-1}), s_{pump} is the pumping speed (m^3·s^{-1}), and p is the desired vacuum pressure (Pa). The pumping speed required to compensate for the gas emanating from the fingerprint and to maintain a partial pressure of $p = 10^{-5}$Pa is

$$s_{pump} = \frac{q_{gas}}{p} = \frac{1 \cdot 10^{-6}\ \mathrm{Pa} \cdot \mathrm{m}^3 \cdot s^{-1}}{1 \cdot 10^{-5}\ \mathrm{Pa}} = 10^{-1}\mathrm{m}^3 \cdot \mathrm{s}^{-1} = 100\,\mathrm{liter} \cdot \mathrm{s}^{-1}. \tag{4.29}$$

It can take several hours to remove the extra partial pressure, not including the actual pumpdown of the vacuum chamber. Figure 4.20 compares two pumpdown curves using the same equipment, one with and one without a fingerprint inside the chamber. The clean chamber without a fingerprint reaches the target pressure of $1.33 \cdot 10^{-5}$ Pa (10^{-7} torr) after 9 h, while the chamber contaminated with a fingerprint will need several more hours to achieve that target pressure. This emphasizes the importance of strict procedures for the assembly and handling of vacuum systems, including the use of vacuum-compatible gloves. □

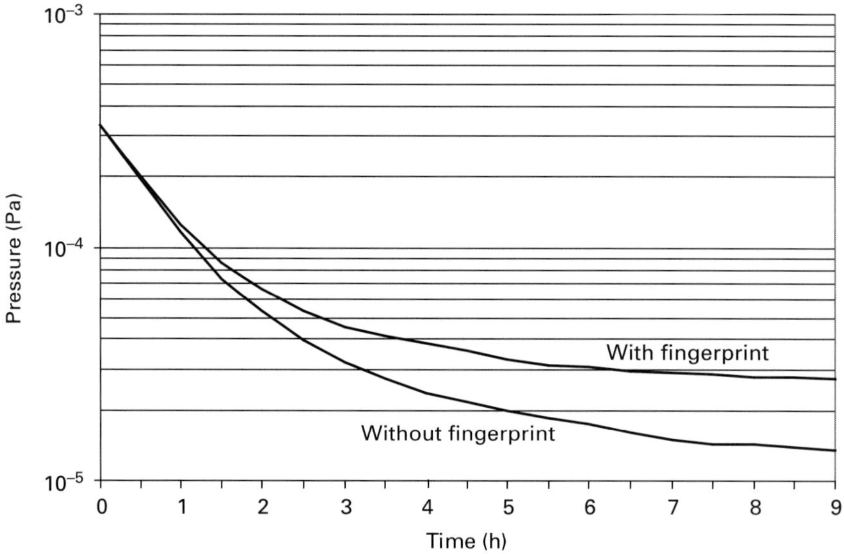

Figure 4.20 Pumpdown curves with and without fingerprint contamination, after Danielson (2001a).

coolants, cleaning solvents, fingerprints, and particles. Both mechanical and chemical cleaning may be necessary, depending on the part and the required vacuum quality (Halliday, 1987b). All personnel must wear neoprene gloves and vapor protection. The compatibility of cleaning solvents with the vacuum system and process environment must be assured. For example, newly machined aluminum can react with toxic and flammable solvents like trichloroethane and trichloroethylene. Non-toxic and non-flammable solvents that evaporate fast and leave no residue are recommended. On-site inspections at vendor facilities are suggested on a regular basis to verify compliance with agreed procedures.

The following cleaning procedure is generally accepted for parts and components. Its point of completion depends on the required vacuum quality: for low vacuum the procedure is completed after Step 3, for medium vacuum after Step 4, and for high vacuum after Step 5. For ultra-high vacuum the entire procedure must be completed.

Step 1. Remove all O-rings and gaskets and wipe them with a lint-free tissue. Volatile solvents should not be used, because they may be absorbed and then outgas in vacuum. O-ring grooves are cleaned with a solvent, and any solvent traces are removed.

Step 2. Remove gross contamination such as cutting oils, flux, and grease with careful mechanical cleaning and wiping, followed by a water rinse.

Step 3. Remove any grease with a solvent, then rinse with tap water and dry with clean, warm air. This completes the cleaning for low vacuum. For medium vacuum continue to Step 4.

Step 4. Rinse with deionized water or clean solvent. A vapor degrease is optional. This completes the cleaning for medium vacuum. For high vacuum continue to Step 5.

Step 5. Clean metal parts for high vacuum using the following steps: (a) electropolish; (b) immerse in detergent under ultrasonic agitation; (c) rinse with tap water; (d) rinse with deionized water; (e) dry in vacuum furnace; (f) install and vacuum-bake at 200 °C. This completes the cleaning for high vacuum. For ultra-high vacuum skip item (f) and continue to Step 6.

Step 6. Clean the parts for ultra-high vacuum using: (a) in-situ vacuum baking at 200 to 400 °C in the target vacuum chamber; (b) further outgassing during operation.

A vacuum furnace should be used for drying. Clean, warm air from a heat gun is an alternative.

4.8.2 Assembly and handling

Vacuum robots must be assembled in a cleanroom environment of ISO Class 10000 or better, with clean work areas and proper cleanroom ventilation. Personnel must wear proper cleanroom attire, including lint-free gloves, hairnets, gowns, and shoe covers. This is recommended for all cleanroom robots, but is imperative for vacuum robots. In particular, surfaces that will be exposed to vacuum must be protected from contamination and from damage such as scratches. The impact of a single fingerprint in vacuum is emphasized in Example 4.6. The industry standard SEMI E49.6-1103 provides guidelines for the assembly and testing of stainless steel systems.

4.8.3 Removing gas loads

The removal of gas loads from parts and components with an appropriate treatment before their use in vacuum usually reduces outgassing by orders of magnitude. This is true even for excellent vacuum materials such as stainless steel and aluminum. See also Table 4.6. The robot design process must include a review of all materials that will be exposed to vacuum. Special attention should be given to the outgassing properties of elastomers and plastics. For example, Viton® O-rings can contain traces of unreacted monomers and low-vapor-pressure plasticizers, which can impact a vacuum manufacturing process in two ways:

- They desorb over weeks or months, adding partial pressure to the vacuum environment.
- They re-condense in the vacuum chamber and contaminate the process and the product.

It is an economic necessity to maximize the productive 'uptime' of a tool, and any extended pumpdown time necessary to compensate for partial pressures from contaminants reduces the productivity of a process tool. Elevated temperatures change the vapor pressure of contaminants (see also Sections 4.7.1 and 4.7.2). Several methods are available for removing gas loads from robot parts before and after installation in a vacuum chamber:

- Vacuum baking
- Vacuum treatment at room temperature
- Atmospheric baking, for example with an infrared lamp.

During vacuum heating the robot parts or assemblies are heated under vacuum for an extended period of time. Vacuum baking forces the accelerated outgassing of potential gas loads and other contaminants. Vacuum-baked parts should not be cleaned again because they may absorb cleaning solvents (Danielson, 2004).

A treatment of parts or subassemblies may include all or some of the following steps, depending on the required vacuum quality:

- Flush the chamber with a hot gas, for example nitrogen, to remove surface contaminants.
- Bake the chamber with infrared lights. (This may leave some untreated cold spots.)
- Desiccate or bake the chamber for an extended time period under low vacuum. Elastomer O-rings can be baked at temperatures up to 150 °C.
- Ultra-high-vacuum systems should be vacuum-baked under vacuum pressure at 200 to 400 °C. Low vacuum is usually sufficient. Metal gaskets can be vacuum-baked at up to 450 °C.
- Electroslag remelting (ESR) is a second refining process that can be applied during production in order to further purify the metal, reduce non-metallic inclusions and voids, and yield a finer, more uniform grain structure.

Example 4.7: pumpdown of a vacuum chamber

New vacuum chambers usually contain residual gas loads, mostly water vapor, but also various contaminants from cleaning agents. The installation of a new vacuum system on the manufacturing floor must include treatments that remove these gas loads. Figure 4.21 compares two pressure curves measured during the pumpdown of the same vacuum chamber, before and after vacuum baking. Note the improvement in chamber performance by two orders of magnitude after the treatment. □

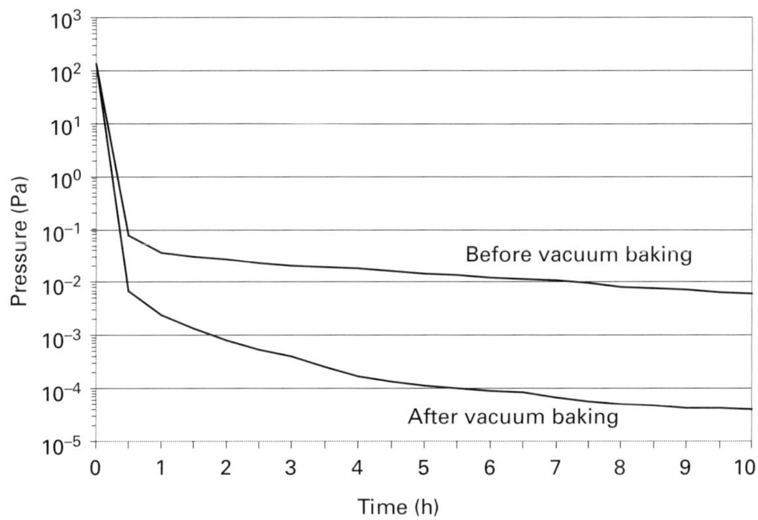

Figure 4.21 Pumpdown curves with and without vacuum baking, after Danielson (2005).

Example 4.8: specification and design concept for a vacuum robot

This example summarizes the product specification and outlines the resulting mechanical design (overview drawings) for the VR300 vacuum robotic system for handling 300 mm wafers in semiconductor manufacturing. The VR300 system includes robot, end-effector, and robot controller. (The controller specifications are not listed.)

The example also highlights the economic impact of a robot's wafer throughput. We assume that a 'wafer swap' takes 8 s (replacing a processed wafer with a new wafer) and that the process tool and the VR300 system operate uninterrupted 24 hours per day, 7 days per week. Therefore 450 wafer swaps per hour are completed, which does not include the time needed for the manufacturing process. If the wafer swap time can be reduced to 7 s the throughput increases by 64 wafer swaps per hour to 514, resulting in 560 640 additional wafers per year. With a hypothetical profit of \$1000 per wafer, the additional annual revenue is \$5 606 400.

Product specification

The vacuum robot will operate in a vacuum cluster tool for wafer processing tools. It will be placed at the center of a six-, seven-, or eight-sided vacuum chamber. The robot work envelope must be sufficient for wafer handling in these environments. The maximum arm extension (with end-effector) is critical. The interfaces of the VR300 robot (electrical, mechanical interface of top flange and wrist, end-effector) with the vacuum cluster tool must comply with the industry standards SEMI E20-0697, SEMI E22.1-1296, and SEMI E21.1-1296. The VR300 robot specification is summarized in Table 4.9.

Wafer throughput is a critical parameter. The VR300 robot must not limit the throughput of the manufacturing process.

Mechanical robot design

The product specification in Table 4.9 was realized with the dual-arm vacuum robot shown in the following figures. Refer to Figure 4.1 for an illustration of a vacuum robot/chamber configuration. The two fully independent arms operate in vacuum. The robot has five axes of motion, described using a cylindrical coordinate system. Each 'frog-leg' arm utilizes two axes of motion, one for the left and one for the right linkage of the closed kinematic chain. The fifth axis is for the vertical axis of motion, common to both arms. Figure 4.22 shows a three-dimensional view of the vacuum robot. Note the identical arms and end-effectors. Figure 4.23 (top) is a top view of the robot with both arms fully retracted. The curved arm shape allows the end-effector to be retracted to the vertical shaft assembly. The minimum sweep radius is smaller than originally specified. Figure 4.23 (center) shows a top view of the robot with both arms fully extended. The maximum extension from the robot center (the *Z*-axis) to the wafer center is 970 mm, more than originally specified. Figure 4.23 (bottom) shows a side view of the robot with both arms fully extended. The vertical gap between the outer links and the gap between the end-effectors of both arms are minimized such that both arms can reach through the narrow openings of load locks. Refer to the SEMI standards E22.1-1296 and E21.1-1296. The minimum sweep radius and the vertical motion range are constrained by the dimensions of the vacuum cluster tool. The four axes of motion for the arms are transferred across the dynamic vacuum barrier into the vacuum. Magnetic feedthrough technology (Section 4.4.1) is used. One single-axis feedthrough is used per axis. The complexity of a single feedthrough with four concentric rotating shafts would pose a reliability risk and would be cost-prohibitive. □

4.9 Applicable and related standards

Several industry standards and guidelines apply directly or indirectly to the robots discussed in this chapter. The following list provides a selection. SEMI standards are

Table 4.9. VR300 robotic system specification (Example 4.8).

Feature	Specification
Wafer size	300 mm
Payload capacity (end-effector and wafer)	1.0 kg (9.81 N)
Mounting method	Top flange
Axes of motion (maximum)	5
Weight (maximum)	45 kg
Vacuum barrier (base pressure in chamber)	10^{-4} Pa ($7.5 \cdot 10^{-7}$ torr)
Vacuum integrity (maximum leak rate)	10^{-7} atm·cm^3·s^{-1} (helium)
Materials exposed to vacuum	Aluminum, stainless steel, Vespel®, Viton®
Motion envelope (cylindrical coordinate frame)	Vertical 50 mm (maximum), horizontal 968 mm radius (minimum)
Sweep radius (minimum)	415 mm
Temperature (maximum)	130 °C intermittent (arm and end-effector) 60 °C for continuous operation (robot base)
Positional repeatability (3 sigma)	Vertical 0.1 mm, horizontal 0.04 mm
Wafer throughput ('wafer swaps')	200 wafers per hour
End-effector	300 mm wafers; off-the-shelf product preferred; custom designs are acceptable
Interfaces and interoperability	Per standards SEMI E20-0697, SEMI E22.1-1296, SEMI E21.1-1296
Software commands	Command set and structure similar to comparable 300 mm vacuum robots

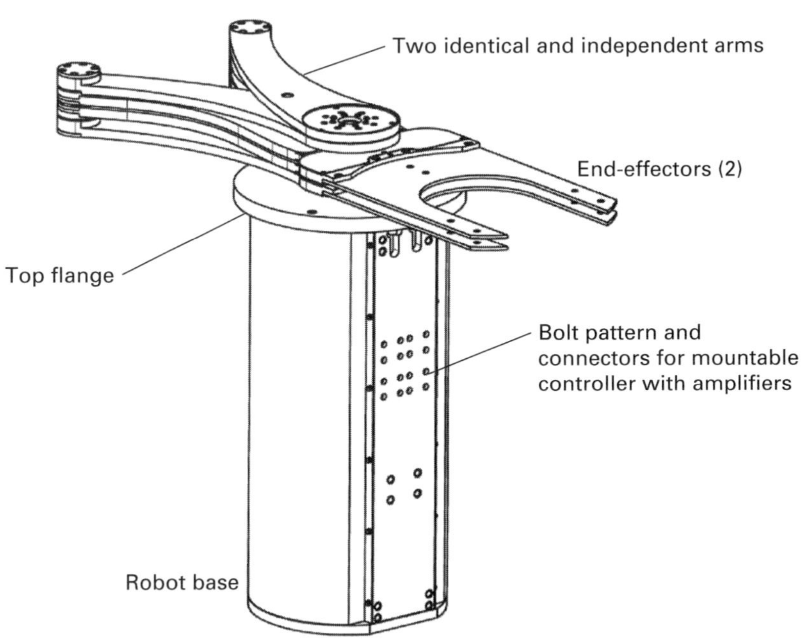

Figure 4.22 3D view of the vacuum robot in Example 4.8. Source: Cymechs Corp.

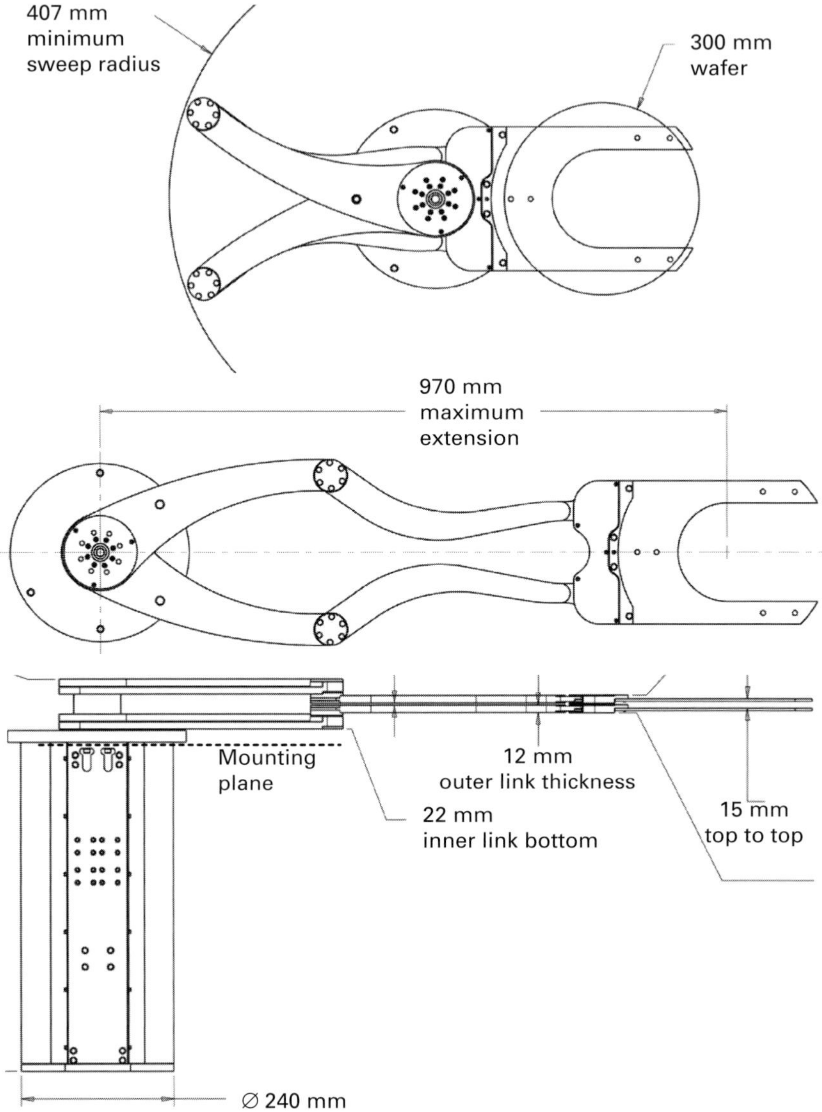

Figure 4.23 Views of the vacuum robot in Example 4.8. Source: Cymechs Corp.

published by Semiconductor Equipment and Materials International (SEMI). Contact information for EMI is listed in Appendix B.

ASTM E 595-77/84/90, Total Mass Loss (TML) and Collected Volatile Condensable Materials (CVCM) from Outgassing in a Vacuum Environment.

SEMI E20-0697, Cluster Tool Module Interface: electrical power and emergency off standard.

SEMI E22.1-1296, Cluster Tool Module Interface 300 mm: Transport module end-effector exclusion volume standard.

SEMI E21.1-1296, Cluster Tool Module Interface 300 mm: mechanical interface and wafer transport standard.

SEMI E49.6-1103, Guide for subsystem assembly and testing procedures – stainless steel systems.

SEMI E106-1104, Overview guide to SEMI standards for physical interfaces and carriers for 300 mm wafers.

SEMI F51-0200, Guide for Elastomeric Sealing Technology.

5 Kinematics

The kinematics model of robots is used to study the geometry of robot motion, in particular the relationship between the end-effector pose (position and orientation) and the joint displacements required to attain a desired end-effector pose. The mechanical structure of industrial robots is an open or closed kinematic chain of links connected by joints. The kinematics model is a system of mathematical equations, usually nonlinear in nature and quite complex for certain robot types. The forces and torques driving a robot manipulator are computed with a dynamical model.

5.1 Joint space, operational space, and workspace

Industrial robots perform their tasks with an end-effector that handles or manipulates a work piece, or applies a tool. Task execution requires computation of the desired end-effector pose, consisting of position $\boldsymbol{p}$ and orientation angle ϕ along the entire end-effector trajectory $\boldsymbol{x}(t)$. $\boldsymbol{p}$ and ϕ become functions of time t, $\boldsymbol{p}(t)$ and $\phi(t)$:

$$\boldsymbol{x}(t) = \begin{bmatrix} x_1(t) \\ \dots \\ x_m(t) \end{bmatrix} = \begin{bmatrix} \boldsymbol{p}(t) \\ \phi(t) \end{bmatrix}, \ \boldsymbol{x} \in \mathbb{R}^m. \tag{5.1}$$

The operational space has m dimensions. A three-dimensional workspace ($m=3$) is typical. The representation of $\boldsymbol{x}(t)$ in Equation (5.1) allows the specification of a robot task with inherently independent parameters. The robot task and the associated $\boldsymbol{x}(t)$ are specified in operational space, which is determined by a robot's kinematic structure. Many robots use Cartesian, polar, or cylindrical coordinate frames. Substrate-handling robots for electronics manufacturing often use cylindrical coordinates, as discussed in Chapters 3 and 4.

The trajectory in joint space is defined as

$$\boldsymbol{q}(t) = \begin{bmatrix} q_1(t) \\ \dots \\ q_n(t) \end{bmatrix}, \ \boldsymbol{q} \in \mathbb{R}^n. \tag{5.2}$$

The robot joint space has n dimensions, one dimension per joint. $\boldsymbol{q}$ is the vector of n joint displacements, or 'generalized joint variables': joint angles for revolute joints and joint positions for prismatic joints. The set of joint displacements is also called 'arm

Example 5.1: three-link planar robot

Consider the planar three-link robot arm illustrated in Figure 5.1. The kinematic structure has three revolute joints, therefore the generalized joint variables are the joint angles $q_i = \theta_i,\ i = 1, 2, 3$. The figure illustrates and relates the kinematic structure of this arm with its joint space, operational space, and workspace. The joint space is three-dimensional, while the operational space is two-dimensional, thus $m = 2$ and $n = 3$ in Equations (5.1) and (5.2), respectively.

Figure 5.1a shows the kinematic structure with three links of unit length, $l_1 = l_2 = l_3 = 1$, and three revolute joints. The robot arm is planar, thus the operational space is described by a two-dimensional Cartesian coordinate frame, the fixed reference frame O_0-x_0y_0. The first joint with joint angle θ_1 coincides with the origin O_0. Links 1 and 2 have joint angles and coordinate frames θ_1, O_1-x_1y_1 and θ_2, O_2-x_2y_2, respectively. The third link is the end-effector with its local coordinate frame O_3-x_3y_3 and joint angle θ_3. The local coordinate frame of a link is 'attached' to the joint that belongs to the next link. The three joint angles combined determine the end-effector position and orientation. The joint space for this particular kinematic structure is shown in Figure 5.1b. For this example the joints are constrained to angles $0 \le \theta_i \le \pi/2, i = 1, 2, 3$; therefore the admissible region in joint space is the cube indicated in the figure. The associated workspace is the region in the operational space that the robot can reach (Figure 5.1c). The workspace boundary is shown by the dashed lines. The origin O_3 at the end-effector tip determines the workspace boundary. The dashed boundary sections result from θ_1, θ_2 and θ_3 displacements. The path from A to B is discussed in Example 5.2. □

configuration.' The trajectory $\boldsymbol{q}(t)$ is important because robot motion is controlled in joint space by adjusting $\boldsymbol{q}$ such that the desired end-effector pose in operational space is achieved over time.

The workspace of a robot is a subset of the operational space and is limited by the mechanical joint limits. A workspace is described by all possible end-effector positions when all joints travel through their respective motion ranges. The literature sometimes distinguishes between reachable and dexterous workspace. A dexterous workspace has boundaries where the end-effector still has all degrees of freedom for achieving a commanded orientation. As such, the dexterous workspace is a subset of the reachable workspace (Sciavicco and Siciliano, 1996).

An open-ended kinematic structure like the planar arm in Figure 5.1 is called an open kinematic chain and is the fundamental structure of industrial robots. Closed kinematic chains are also, but less frequently, used. The conventional graphical representations of kinematic structures are kinematic skeleton structures, simple illustrations of joints connected by lines. Figure 5.2a and Figure 5.2b illustrate revolute and prismatic joints, respectively. Revolute joints allow rotation; prismatic joints allow translation. Figure 5.2c shows the skeleton structure of a SCARA-type robot, an open kinematic

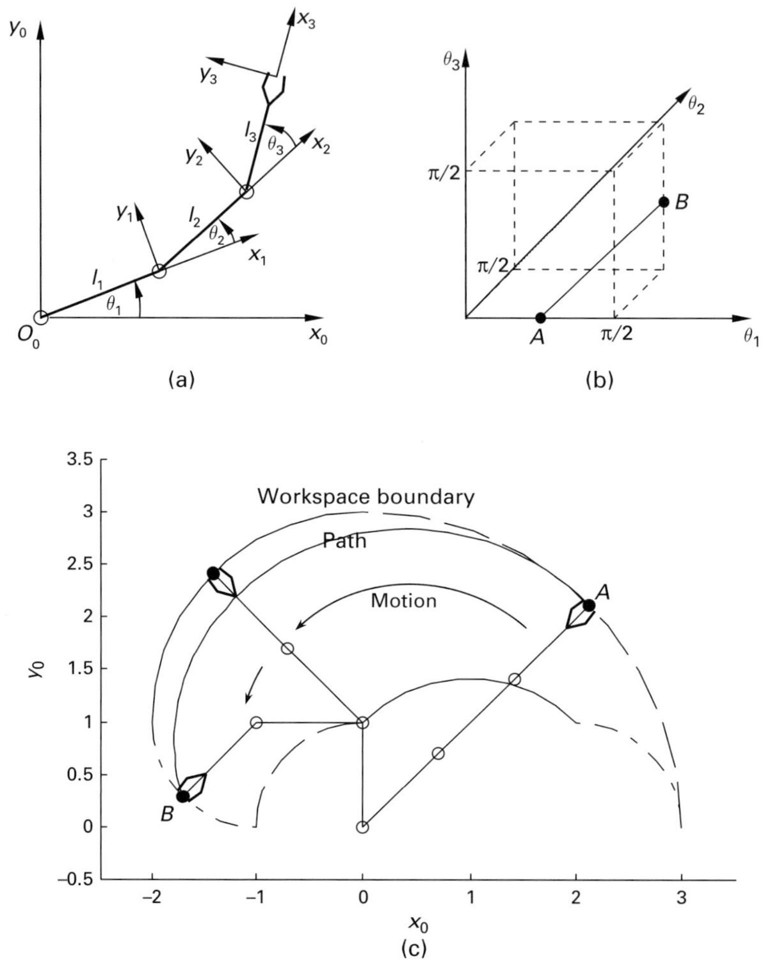

Figure 5.1 Planar three-link robot arm: (a) kinematic structure, (b) joint space, motion path A to B, (c) workspace with boundaries and path A to B (see also Example 5.2).

chain with the planar three-link arm in Figure 5.1 plus one vertical axis. This robot type is regularly used in electronics manufacturing, particularly in the semiconductor industry, and is analyzed in detail throughout this book. Figure 5.2d shows a closed kinematic chain based on two planar three-link arms. This skeleton structure, a 'frog-leg arm,' forms the basis for popular vacuum robot models in semiconductor manufacturing.

The following section presents common kinematic structures of industrial robots.

5.2 Kinematic robot structures

Robots are categorized by their kinematic structure and the coordinate frame used to describe the robot's workspace. The following robot types are common and are defined in the international ISO standard 8373.

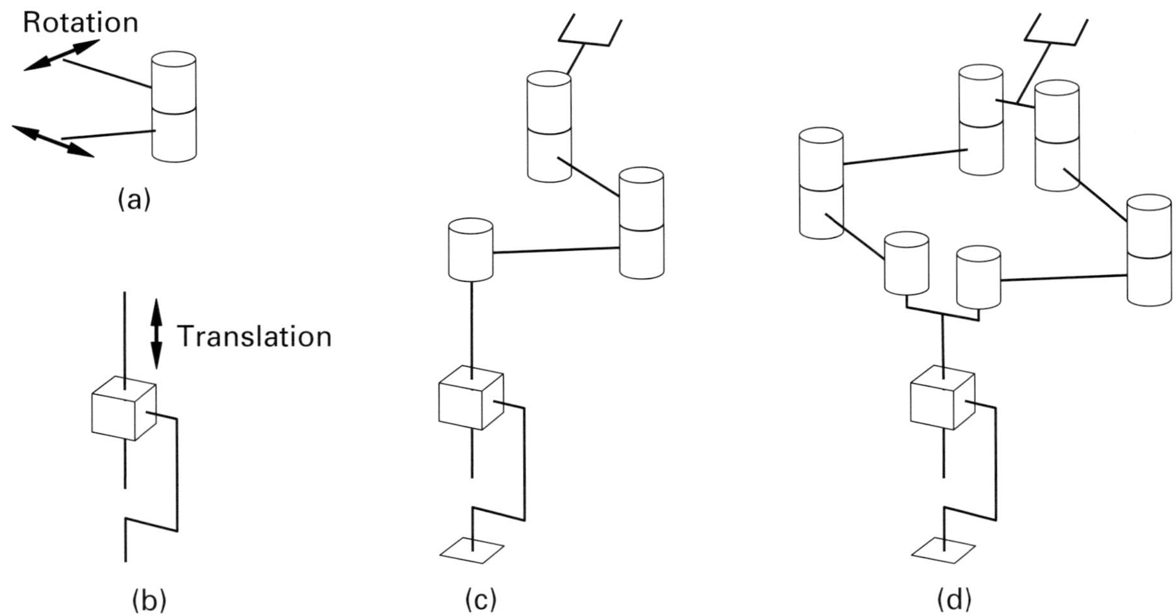

Figure 5.2 Kinematic skeleton structures: (a) revolute joint, (b) prismatic joint, (c) open kinematic chain, (d) closed kinematic chain.

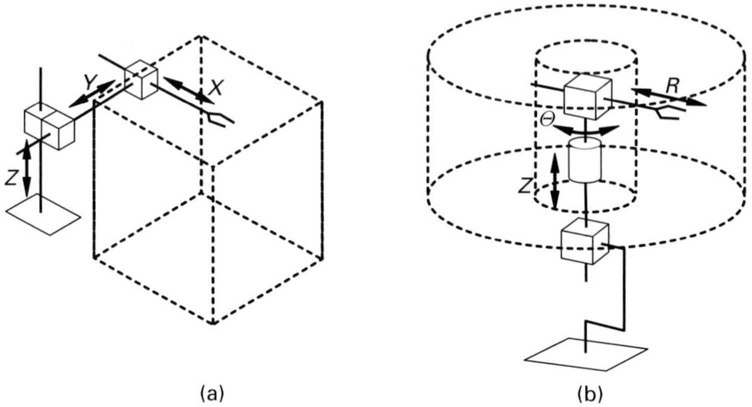

Figure 5.3 Skeleton structure and workspace: (a) Cartesian, (b) cylindrical robot.

5.2.1 Rectangular or Cartesian robots

A rectangular or Cartesian robot is defined in ISO 8373:1994, No. 3.15.1 as a "robot whose arm has three prismatic joints, whose axes are coincident with a Cartesian coordinate system." The end-effector may use additional joints. One example is gantry robots, used in applications that require accurate and repeatable transfer and positioning of heavy loads. They are mounted either to the ceiling or to a rigid frame to support the robot. This robot type is constrained to move linearly and parallel to the axes of a Cartesian coordinate frame. Cantilevered Cartesian robots include arm, carriage, column, and a base (Figure 5.3a), are accurate and repeatable, and are easy to program. Arm rigidity can generally be achieved with a limited maximum arm extension.

5.2.2 Cylindrical robots

A cylindrical robot is defined in ISO 8373:1994, No. 3.15.2 as a "robot whose arm has at least one rotary and at least one prismatic joint and whose axes form a cylindrical coordinate system." See Figure 5.3b. The end-effector may use additional joints. The workspace is a cylinder, or part of a cylinder. A typical kinematic structure includes the base, a vertical column, and a carriage that moves along the vertical column using a prismatic joint. The vertical column is mounted to the base with a revolute joint that provides robot rotation. Some robot designs provide more than 360° of rotation. The horizontal arm, mounted to the carriage, extends and retracts radially. The reachable workspace may be 'hollow,' due to a limited arm retraction, as is shown in Figure 5.3b. One version of a cylindrical robot has SCARA-type, jointed arms with pivoted links. Cylindrical robots are widely used in electronics manufacturing, and this book will focus on these and on the SCARA robots discussed below.

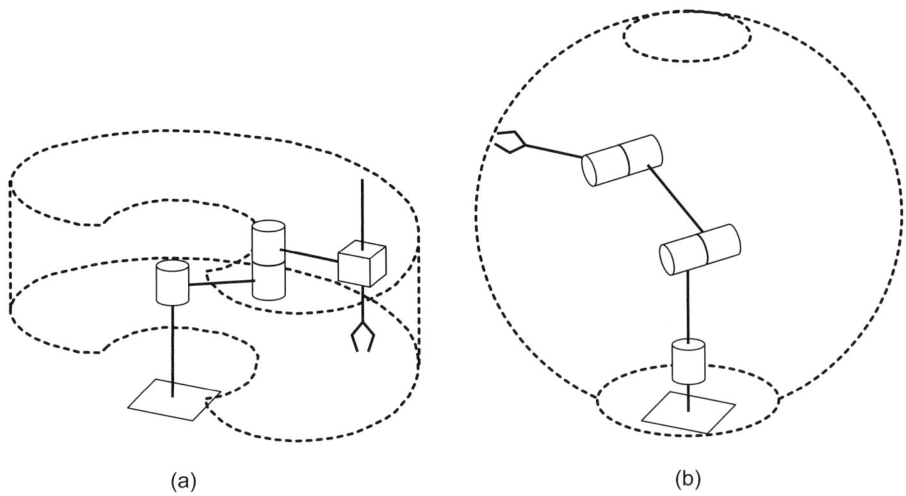

Figure 5.4 Kinematic structure and workspace: (a) SCARA robot, (b) articulated robot.

5.2.3 SCARA robots

A SCARA robot is defined in ISO 8373:1994, No. 3.15.6, as a "robot which has two parallel revolute joints to provide compliance in a selected plane," and may be considered a special case of a cylindrical robot. The term 'SCARA' stands for 'selective compliance arm for robot assembly'. (Other interpretations include 'selective compliance articulated robot arm'.) 'Compliance' in robotics refers to the elasticity of a manipulator, and is defined as the "flexible behavior of a robot or any associated tool in response to external forces exerted on it. When the behavior is independent of sensory feedback, it is passive compliance; if not, it is active compliance" (ISO 8373 No. 5.3.7).

The motion of four-axis SCARA robots is similar to that of a horizontal and planar three-link arm configuration (shoulder, elbow, wrist) with the addition of an up-and-down component (Figure 5.4a). Typical applications for SCARA robots are assembly, packaging, sorting, light-duty drilling, and screwdriving. If mechanical constraints limit the arm motion to radial paths it can also be classified as a cylindrical, SCARA-type robot, which is used in manufacturing tools.

5.2.4 Anthropomorphic or articulated robots

ISO 8373:1994, No. 3.15.5, defines an anthropomorphic robot as a "robot whose arm has three revolute joints." See Figure 5.4b. The end-effector may use additional joints.

5.3 Mathematical foundations

This section reviews the mathematical foundations needed for creating kinematic models of robots.

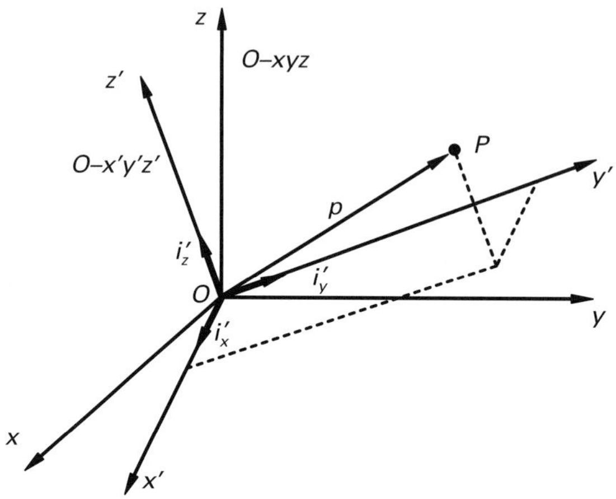

Figure 5.5 Rotational transform of point *P*.

5.3.1 Rigid body in three-dimensional space

Robot manipulators are modeled as a chain of rigid bodies in the m-dimensional operational space, for example a Cartesian or cylindrical coordinate frame. A rigid body is completely described by its position and orientation, that is, a six-dimensional state vector.

Position (translation)

Position can be modeled with a translation, a mathematical operation. Consider a fixed Cartesian reference frame *O-xyz* with origin *O* and axes *x*, *y*, *z*, and the vector $\boldsymbol{p}$ from *O* to an arbitrary point *P* (Figure 5.5). Vector $\boldsymbol{p}$ determines the position of *P* and is defined by the three coordinates p_x, p_y, p_z along the three axes with unit vectors $\boldsymbol{i}_x, \boldsymbol{i}_y, \boldsymbol{i}_z$:

$$\boldsymbol{p} = p_x \cdot \boldsymbol{i}_x + p_y \cdot \boldsymbol{i}_y + p_z \cdot \boldsymbol{i}_z = \begin{bmatrix} \boldsymbol{i}_x & \boldsymbol{i}_y & \boldsymbol{i}_z \end{bmatrix} \cdot \begin{bmatrix} p_x \\ p_y \\ p_z \end{bmatrix} = \boldsymbol{I} \cdot \begin{bmatrix} p_x \\ p_y \\ p_z \end{bmatrix} = \begin{bmatrix} p_x \\ p_y \\ p_z \end{bmatrix}. \tag{5.3}$$

Bold symbols indicate a vector or matrix. $\boldsymbol{I}$ is the identity matrix. $\boldsymbol{p}$ performs a translation from *O* to *P*, the position of the rigid body. Here the vector $\boldsymbol{p}$ positions the reference point of a robot arm link at *P* (the joint at the end of the link) with respect to the reference frame *O-xyz*.

Orientation (rotation)

Body orientation can be modeled with a rotation. A coordinate frame *O-x′y′z′* is attached to the rigid body. As the body rotates, so does *O-x′y′z′*, with respect to *O-xyz*. *P* has different coordinates in each frame. The mathematical tool for switching between reference frames is a coordinate transform. Figure 5.5 shows point *P* in both the fixed ('world') frame *O-xyz* and the rotated frame *O-x′y′z′*. Both frames coincide at the origin *O*. Vector $\boldsymbol{p}'$ is transformed from *O-x′y′z′* to $\boldsymbol{p}$ in *O-xyz* using the rotation matrix $\boldsymbol{R} = \begin{bmatrix} \boldsymbol{i}'_x & \boldsymbol{i}'_y & \boldsymbol{i}'_z \end{bmatrix}$:

$$p = p'_x \cdot i'_x + p'_y \cdot i'_y + p'_z \cdot i'_z = \begin{bmatrix} i'_x & i'_y & i'_z \end{bmatrix} \cdot \begin{bmatrix} p'_x \\ p'_y \\ p'_z \end{bmatrix} = R \cdot p'. \tag{5.4}$$

The rotation matrix is orthonormal in the Cartesian reference frame with the orthogonal (linearly independent) unit vectors, thus

$$R^T R = I, R^{-1} = R^T. \tag{5.5}$$

The inverse coordinate transform that maps p from the fixed world frame into $O\text{-}x'y'z'$ is thus given by

$$p' = R^T \cdot p. \tag{5.6}$$

The rotation matrix is derived below.

5.3.2 Rotation matrix

The frame $O\text{-}x'y'z'$ can be rotated in the world frame $O\text{-}xyz$ with three elementary rotations, that is, one rotation about each of the axes x, y, z. For example, $O\text{-}x'y'z'$ is rotated about the z'-axis by the angle α using the rotation matrix R_z. Note that $|i'| = 1$. The rotating x/y plane is illustrated in Figure 5.6 (left). The figure also shows how the coordinates x', y' are mapped from $O\text{-}x'y'z'$ to $O\text{-}xyz$. This operation is conveniently computed with R_z:

$$R_z = \begin{bmatrix} i'_x & i'_y & i'_z \end{bmatrix} = \begin{bmatrix} \cos(\alpha) & -\sin(\alpha) & 0 \\ \sin(\alpha) & \cos(\alpha) & 0 \\ 0 & 0 & 1 \end{bmatrix}. \tag{5.7}$$

R_z can be derived with straightforward geometry by transforming P from $O\text{-}xyz$ to $O\text{-}x'y'z'$ (Figure 5.7). Let P^* be the projection of P onto the x/y plane. The rows of R_z are obtained by simple trigonometry. For example, the first row is

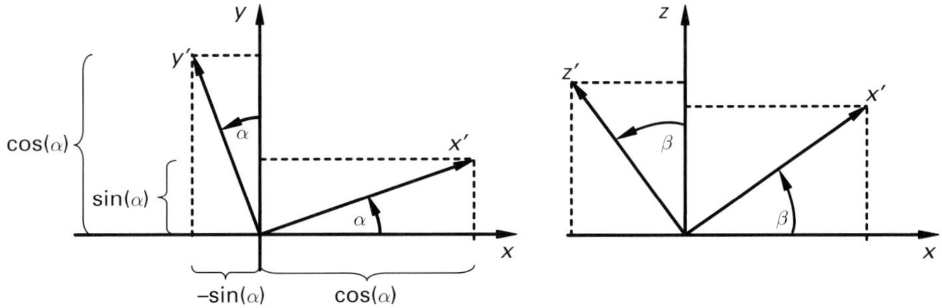

Figure 5.6 Rotation about the z-axis (left) and y-axis (right).

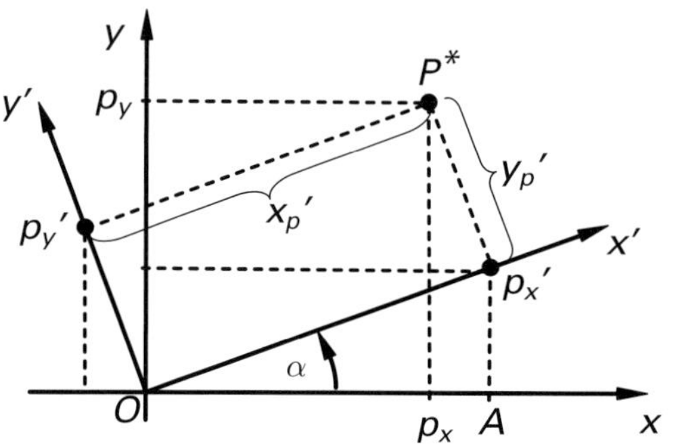

Figure 5.7 Rotation about the z-axis using the rotation matrix $\boldsymbol{R}_z$.

$$\begin{aligned}\overline{Op_x} &= \overline{OA} - \overline{p_x A} \\ &= x_p' \cdot \cos(\alpha) - y_p' \cdot \sin(\alpha) \\ &= \begin{bmatrix} \cos(\alpha) & -\sin(\alpha) & 0 \\ \cdots & \cdots & \cdots \\ \cdots & \cdots & \cdots \end{bmatrix} \cdot \begin{bmatrix} x_p' \\ y_p' \\ z_p' \end{bmatrix}.\end{aligned} \tag{5.8}$$

Rotations about the y-axis by the angle β are computed with $\boldsymbol{R}_y$ (Figure 5.6, right), and rotations about the x-axis by the angle γ are computed with $\boldsymbol{R}_x$:

$$\boldsymbol{R}_y = \begin{bmatrix} \cos(\beta) & 0 & -\sin(\beta) \\ 0 & 1 & 0 \\ \sin(\beta) & 0 & \cos(\beta) \end{bmatrix}, \quad \boldsymbol{R}_x = \begin{bmatrix} 1 & 0 & 0 \\ 0 & \cos(\gamma) & -\sin(\gamma) \\ 0 & \sin(\gamma) & \cos(\gamma) \end{bmatrix}. \tag{5.9}$$

The three rotation matrices can be multiplied to obtain the general rotation matrix $\boldsymbol{R}$:

$$\boldsymbol{p} = \boldsymbol{R} \cdot \boldsymbol{p}' = \boldsymbol{R}_x \cdot \boldsymbol{R}_y \cdot \boldsymbol{R}_z \cdot \boldsymbol{p}', \tag{5.10}$$

where $\boldsymbol{p}$ and $\boldsymbol{p}'$ are the vectors pointing to P^* in O-xyz and O-$x'y'z'$, respectively.

Although $\boldsymbol{R}$ is a convenient mathematical operator for describing frame orientation, it is redundant (its nine elements are not independent) and therefore is not efficient for software implementation. Three independent parameters suffice to represent the rotation of a rigid body in a Cartesian coordinate frame. Two such minimal representations are the Euler angles and the roll, pitch, and yaw angles.

5.3.3 Homogenous coordinate transforms

The above discussion has described position and orientation separately. A coordinate transform that describes the end-effector pose $\boldsymbol{p}$ by combining position and orientation is

$$\boldsymbol{p} = \boldsymbol{p}_0 + \boldsymbol{R} \cdot \boldsymbol{p}'. \tag{5.11}$$

This is illustrated in Figure 5.8. The rotating frame O'-$x'y'z'$ with origin O' is attached to vector $\boldsymbol{p}_0$. The inverse transform is

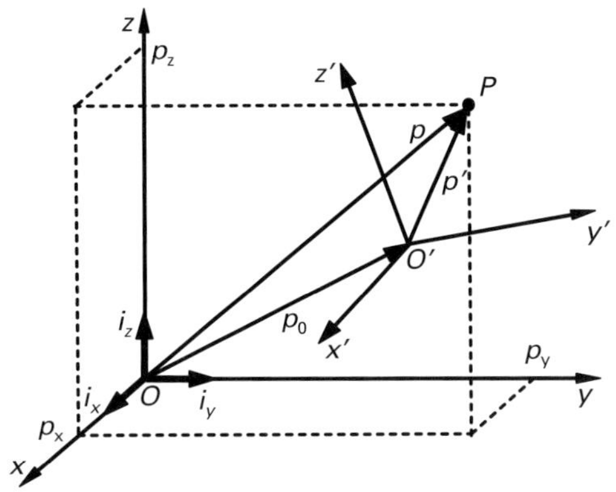

Figure 5.8 Coordinate transform of point P to the world frame O-xyz.

$$\boldsymbol{p}' = \boldsymbol{R}^T \cdot (-\boldsymbol{p}_0 + \boldsymbol{p}). \tag{5.12}$$

In practice this is used to transform a work piece from end-effector coordinates to the robot's coordinates.

Each arm link in the open kinematic chain of industrial robots is described by its position and orientation with respect to the local coordinate frame of the previous arm link, and eventually with respect to the fixed reference frame of the robot base. A compact, 'homogenous' coordinate transform combines both translation and orientation. This is achieved by rewriting Equation (5.11) as

$$\tilde{\boldsymbol{p}} = \boldsymbol{A} \cdot \tilde{\boldsymbol{p}}' = \begin{bmatrix} \boldsymbol{R} & \boldsymbol{p}_0 \\ 0^T & 1 \end{bmatrix} \cdot \begin{bmatrix} \boldsymbol{p}' \\ 1 \end{bmatrix}, \tag{5.13}$$

with $\mathbf{0}^T = [0\ 0\ \ldots\ 0]$. The transformation matrix $\boldsymbol{A}$ is not orthonormal, so $\boldsymbol{A}^{-1} \neq \boldsymbol{A}^T$. The inverse homogenous transform is given by

$$\tilde{\boldsymbol{p}}' = \boldsymbol{A}^{-1} \cdot \tilde{\boldsymbol{p}} = \begin{bmatrix} \boldsymbol{R}^T & -\boldsymbol{R}^T\boldsymbol{p}_0 \\ \mathbf{0}^T & 1 \end{bmatrix} \cdot \begin{bmatrix} \boldsymbol{p} \\ 1 \end{bmatrix}. \tag{5.14}$$

Equation (5.13) is used to describe both position and orientation of links $0 \ldots n$ in a kinematic chain. The robot base is link 0, with the fixed coordinate frame O_0-$x_0y_0z_0$. Equation (5.14) is used to describe the pose of link n, the end-effector with frame O_n-$x_ny_nz_n$, with respect to O_0-$x_0y_0z_0$. This is needed because the end-effector performs robot tasks that are specified in O_0-$x_0y_0z_0$. The end-effector pose with respect to O_0-$x_0y_0z_0$ is obtained by consecutive homogenous transforms from link n backward to link 0:

$$\boldsymbol{p}_0 = \boldsymbol{A}_n^0 \cdot \boldsymbol{p}_n = \boldsymbol{A}_1^0 \cdot \boldsymbol{A}_2^1 \cdot \ldots \cdot \boldsymbol{A}_n^{n-1} \cdot \boldsymbol{p}_n. \tag{5.15}$$

Notation: the matrix $\boldsymbol{A}_n^{n-1}$ transforms the end-effector pose from frame O_n-$x_ny_nz_n$ to frame $O_{n\text{-}1}$-$x_{n\text{-}1}y_{n\text{-}1}z_{n\text{-}1}$.

5.4 Forward kinematics

This section discusses the manipulator forward kinematics, the mathematical model of the robot geometry, and the mapping from joint space to operational space. The robot inverse kinematics map a point in workspace to the associated point in joint space (Asada and Slotine, 1986; Mathia and Saeks, 1994).

5.4.1 Mathematical model

Forward kinematics is used for open chains of links connected by joints. Each joint provides one degree of freedom. The Robotic Industries Association (RIA) defines forward kinematics as "the mathematical relationship which determines the pose of the robot tool coordinate system (TCS) based on the values of the principal joint displacements" (RIA standard R15.07). The TCS is the end-effector coordinate frame.

Forward kinematics of robots with n joints that operate in an m-dimensional space are described with a vector-valued function of the form

$$\boldsymbol{x} = \begin{bmatrix} \boldsymbol{p} \\ \phi \end{bmatrix} = \boldsymbol{f}(\boldsymbol{q}), \quad \boldsymbol{q} \in \mathbb{R}^{n\times 1}, \boldsymbol{x} \in \mathbb{R}^{m\times 1}. \tag{5.16}$$

The function $\boldsymbol{f}: \mathbb{R}^n \mapsto \mathbb{R}^m$ computes the end-effector position $\boldsymbol{x}$ (position $\boldsymbol{p}$ and orientation ϕ) given the joint displacements $\boldsymbol{q}$. $\boldsymbol{f}$ is nonlinear if the kinematic structure includes at least one revolute joint. Only Cartesian robots have purely linear kinematic structures. In the most general case a robot is controllable in all dimensions, that is, the positions p_x, p_y, p_z (from here on, simply x, y, z), and the roll, pitch, and yaw angles φ, ϑ, ψ:

$$\boldsymbol{x} = \begin{bmatrix} x \\ y \\ z \\ \varphi \\ \vartheta \\ \psi \end{bmatrix} = \begin{bmatrix} f_1(\boldsymbol{q}) \\ f_2(\boldsymbol{q}) \\ f_3(\boldsymbol{q}) \\ f_4(\boldsymbol{q}) \\ f_5(\boldsymbol{q}) \\ f_6(\boldsymbol{q}) \end{bmatrix}, \quad \boldsymbol{q} \in \mathbb{R}^{n\times 1}, \ \boldsymbol{x} \in \mathbb{R}^{6\times 1}. \tag{5.17}$$

However, the maximum flexibility provided by such a manipulator results in a complex mechanical design and control problem. Therefore robot designs generally are limited to the minimum number of degrees of freedom required for a given application or range of applications. Two methods are available to establish the forward kinematic for a given robot type: straightforward geometry and transformation matrices. Denavit–Hartenberg notation is commonly used for establishing transformation matrices.

5.4.2 Denavit–Hartenberg notation

The straightforward geometry method is feasible only for fairly simple kinematic structures, but becomes difficult for more complex robot arms. Denavit–Hartenberg (DH) notation is a systematic method for establishing a forward kinematics model. It

Example 5.2: forward kinematics from straightforward geometry

Consider the planar three-link robot arm in Figure 5.1. The forward kinematics that maps the joint angles $\theta = [\theta_1 \theta_2 \theta_3]^T$ to $\boldsymbol{x} = [x\ y\ \phi]^T$ is derived using straightforward geometry. The resulting nonlinear vector-valued function is

$$\boldsymbol{x} = \begin{bmatrix} x \\ y \\ \phi \end{bmatrix} = \begin{bmatrix} l_1 \cdot \cos(\theta_1) + l_2 \cdot \cos(\theta_1 + \theta_2) + l_3 \cdot \cos(\theta_1 + \theta_2 + \theta_3) \\ l_1 \cdot \sin(\theta_1) + l_2 \cdot \sin(\theta_1 + \theta_2) + l_3 \cdot \sin(\theta_1 + \theta_2 + \theta_3) \\ \theta_1 + \theta_2 + \theta_3 \end{bmatrix}. \quad (5.18)$$

Figure 5.1b shows a commanded path in joint space from point $A = [\pi/4\ 0\ 0]^T$ to point $B = [\pi/2\ \pi/2\ \pi/4]^T$. Figure 5.1c shows the resulting path in workspace, computed with Equation (5.18) and described by the end-effector's origin O_3. The figure also shows the arm for the initial and final end-effector poses at A and B. □

guides the assignment of coordinate frames to robot arm links using the minimum number of parameters necessary to completely describe the forward kinematics (Asada and Slotine, 1986; Craig, 1989; Denavit and Hartenberg, 1955; Sciavicco and Siciliano, 1996).

The DH notation for the special case with two adjacent links connected by revolute joints is shown in Figure 5.9. The links are numbered i-1 and i. The i-th link is controlled using the i-th joint. The local coordinate frame O_i-$x_i y_i z_i$ is at the end of link i. The pose of O_i-$x_i y_i z_i$ in the DH notation is obtained using the steps summarized below. The line $B_i O_i$ is the common normal to the joint axes i and i+1. The origin O_i is positioned at the intersection of the joint axis i+1 and $B_i O_i$. The x_i-axis follows the extension of $B_i O_i$. The z_i-axis is along the joint axis i+1. The y_i-axis is selected such that the frame O_i-$x_i y_i z_i$ forms a right-hand coordinate system.

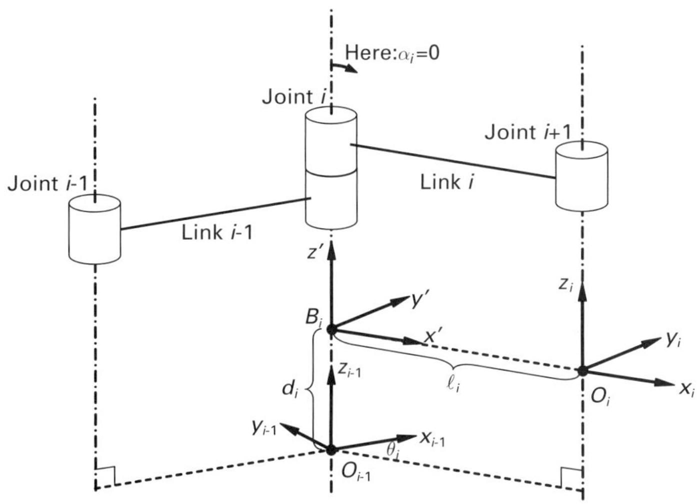

Figure 5.9 DH notation for two adjacent links and joints with parallel axes of rotation.

The relative position and orientation of the two frames is completely determined using four DH parameters:

- l_i: length of the common normal $B_i\ O_i$ (length of link i)
- α_i: tilt angle between joint axis i and the z_i-axis (here: $\alpha_i = 0$)
- d_i: vertical distance between O_{i-1} and B_i
- θ_i: angle between the x_{i-1}-axis and the x_i-axis.

The parameters l_i and α_i depend only on the link geometry and are therefore constants. The planar robot arms with parallel axes of rotation discussed in this book have $\alpha_i = 0$. The controlled parameters are d_i and θ_i. If joint i is a revolute joint the angle θ_i is the controlled parameter. Otherwise, if joint i is a prismatic joint, d_i is the controlled parameter.

With DH notation the kinematic relationship between the links i-1 and i in Figure 5.9 can be described using a four-by-four homogeneous transformation matrix. This is demonstrated below with two transforms: from O_i-$x_iy_iz_i$ backward to the supplemental frame B_i-$x'y'z'$, and from B_i-$x'y'z'$ backward to O_{i-1}-$x_{i-1}y_{i-1}z_{i-1}$. Using the $\boldsymbol{A}$-matrix in Equation (5.13), the transform from the frame O_i-$x_iy_iz_i$ back to the supplemental frame B_i -$x'y'z'$ is

$$\tilde{\boldsymbol{x}}' = \boldsymbol{A}_i^{\text{int}} \cdot \tilde{\boldsymbol{x}}_i = \begin{bmatrix} \boldsymbol{R}_{z,i} & \boldsymbol{x}'_0 \\ \boldsymbol{0}^T & 1 \end{bmatrix} \cdot \begin{bmatrix} \boldsymbol{x}_i \\ 1 \end{bmatrix}. \tag{5.19}$$

For the two planar links the tilt angle α_i is always zero. No tilt rotation is needed. Only a translation by the link length l_i along the x_i-axis applies:

$$\begin{aligned} \begin{bmatrix} \boldsymbol{x}' \\ 1 \end{bmatrix} &= \begin{bmatrix} x' \\ y' \\ z' \\ 1 \end{bmatrix} = \begin{bmatrix} \boldsymbol{0} & \boldsymbol{x}'_0 \\ \boldsymbol{0}^T & 1 \end{bmatrix} \cdot \begin{bmatrix} \boldsymbol{x}_i \\ 1 \end{bmatrix} = \begin{bmatrix} \cos(0) & -\sin(0) & 0 & l_i \\ \sin(0) & \cos(0) & 0 & 0 \\ 0 & 0 & 1 & 0 \\ 0 & 0 & 0 & 1 \end{bmatrix} \cdot \begin{bmatrix} x_i \\ y_i \\ z_i \\ 1 \end{bmatrix} \\ &= \begin{bmatrix} 1 & 0 & 0 & l_i \\ 0 & 1 & 0 & 0 \\ 0 & 0 & 1 & 0 \\ 0 & 0 & 0 & 1 \end{bmatrix} \cdot \begin{bmatrix} x_i \\ y_i \\ z_i \\ 1 \end{bmatrix} = \begin{bmatrix} x_i + l_i \\ y_i \\ z_i \\ 1 \end{bmatrix}. \end{aligned} \tag{5.20}$$

Equation (5.20) gives the position of O_i in the supplemental frame B_i-$x'y'z'$, with $x' = x_i + l_i$. The transform from frame B_i-$x'y'z'$ to O_{i-1}-$x_{i-1}y_{i-1}z_{i-1}$ includes a translation by the vertical offset d_i and a rotation about the z_{i-1}-axis by the angle θ_i :

$$\tilde{\boldsymbol{x}}_{i-1} = \boldsymbol{A}_{\text{int}}^{i-1} \cdot \tilde{\boldsymbol{x}}' = \begin{bmatrix} \boldsymbol{R}_{z,i-1} & \boldsymbol{x}_{i-1,0} \\ \boldsymbol{0}^T & 1 \end{bmatrix} \cdot \begin{bmatrix} \boldsymbol{x}' \\ 1 \end{bmatrix} \tag{5.21}$$

$$\begin{bmatrix} \boldsymbol{x}_{i-1} \\ 1 \end{bmatrix} = \begin{bmatrix} \cos(\theta_i) & -\sin(\theta_i) & 0 & 0 \\ \sin(\theta_i) & \cos(\theta_i) & 0 & 0 \\ 0 & 0 & 1 & d_i \\ 0 & 0 & 0 & 1 \end{bmatrix} \cdot \begin{bmatrix} x_i + l_i \\ y_i \\ z_i \\ 1 \end{bmatrix}$$
$$= \begin{bmatrix} \cos(\theta_i) \cdot (x_i + l_i) - \sin(\theta_i) \cdot y_i \\ \sin(\theta_i) \cdot (x_i + l_i) + \cos(\theta_i) \cdot y_i \\ z_i + d_i \\ 1 \end{bmatrix}. \tag{5.22}$$

The combined transform of Equations (5.20) and (5.22) gives:

$$\tilde{\boldsymbol{x}}_{i-1} = \boldsymbol{A}_{\text{int}}^{i-1} \cdot \boldsymbol{A}_i^{\text{int}} \cdot \tilde{\boldsymbol{x}}_i = \boldsymbol{A}_i^{i-1} \cdot \tilde{\boldsymbol{x}}_i$$
$$= \begin{bmatrix} \cos(\theta_i) & -\sin(\theta_i) & 0 & 0 \\ \sin(\theta_i) & \cos(\theta_i) & 0 & 0 \\ 0 & 0 & 1 & d_i \\ 0 & 0 & 0 & 1 \end{bmatrix} \cdot \begin{bmatrix} 1 & 0 & 0 & l_i \\ 0 & 1 & 0 & 0 \\ 0 & 0 & 1 & 0 \\ 0 & 0 & 0 & 1 \end{bmatrix} \cdot \begin{bmatrix} x_i \\ y_i \\ z_i \\ 1 \end{bmatrix}$$
$$= \begin{bmatrix} \cos(\theta_i) & -\sin(\theta_i) & 0 & \cos(\theta_i) \cdot l_i \\ \sin(\theta_i) & \cos(\theta_i) & 0 & \sin(\theta_i) \cdot l_i \\ 0 & 0 & 1 & d_i \\ 0 & 0 & 0 & 1 \end{bmatrix} \cdot \begin{bmatrix} x_i \\ y_i \\ z_i \\ 1 \end{bmatrix} \tag{5.23}$$
$$= \begin{bmatrix} \cos(\theta_i) \cdot (x_i + l_i) - \sin(\theta_i) \cdot y_i \\ \sin(\theta_i) \cdot (x_i + l_i) + \cos(\theta_i) \cdot y_i \\ z_i + d_i \\ 1 \end{bmatrix}.$$

5.4.3 Generalized coordinates and kinematic equations

It is convenient to express the joint displacements with a generalized joint variable $\boldsymbol{q}_i,\ i = 1, \ldots, n$, where

$$\begin{aligned} q_i &= \theta_i, \text{ joint angle for revolute joint } i \\ q_i &= d_i, \text{ joint position for prismatic joint } i. \end{aligned} \tag{5.24}$$

The homogenous transformation matrix $\boldsymbol{A}_i^{i-1}$ in Equation (5.23) then becomes a function of $\boldsymbol{q}_i$, and the end-effector pose is a function of n joint displacements, each driven by a dedicated actuator. As such, the i-th transformation matrix is a function of the i-th generalized joint variable. Denavit–Hartenberg notation and the generalized coordinates allow construction of the forward kinematics of any open kinematic chain with n links and joints in the form of a single transformation matrix $\boldsymbol{T}_n^0$. Each joint contributes one $\boldsymbol{A}$-matrix:

$$\begin{aligned} \tilde{\boldsymbol{x}}_0 &= \boldsymbol{T}_n^0(\boldsymbol{q}) \cdot \tilde{\boldsymbol{x}}_n,\ \boldsymbol{q} \in \mathbb{R}^n \\ &= A_1^0(q_1) \cdot A_2^1(q_2) \cdot \ldots \cdot A_n^{n-1}(q_n) \cdot \tilde{x}_n. \end{aligned} \tag{5.25}$$

Equation (5.25) is the main result of this section.

The procedure for establishing the forward kinematics using DH notation is summarized below for SCARA-type robots (Figure 5.2c) with n links and n joints. A procedure for the general case can be found in the literature, for example in (Sciavicco and Siciliano, 1996).

Step 1. Identify all joints and number them consecutively $1,\ldots, n$.

Step 2. Label all links consecutively $0,\ldots, n$, beginning with the base link. Link i is moved using joint i.

Step 3. Assign a coordinate frame to each link. The coordinate frame for link i is located at joint $i+1$ at the end of link i. Place the z_i-axis in line with joint axis i+1. The base link (link 0) has the z_0-axis in line with joint axis 1. Place the origin O_0 anywhere on the z_0-axis, preferably at a mechanical limit. Select axes x_0 and y_0 such the frame O_0-$x_0y_0\, z_0$ is a right-handed frame.

Step 4. If the z_i and z_{i-1}-axes are parallel and if joint i is revolute, place O_i such that $d_i =$ 0. If joint i is prismatic then O_i can be anywhere on the z-axis (the joint axis). A convenient reference position for the joint range is desirable, for example, a mechanical limit.

Step 5. Choose axis x_i at a right angle to the z_i-axis with direction from joint i to joint i+1. Select axis y_i such that the frame O_i -$x_iy_iz_i$ is a right-handed frame.

Step 6. For link n (the end-effector) the x_n-axis intersects the joint axis $n-1$ at a right angle. If joint n-1 is revolute, select the z_n-axis parallel to z_{n-1}.

Step 7. For all n links create a table with DH parameters l_i, α_i, d_i, θ_i for links $i = 1,\ldots, n$. Remember that $\alpha_i = 0$ for planar robot arms.

Step 8. Using the DH parameters determine all n homogeneous transformation matrices $\boldsymbol{A}_i^{i-1}(q_i)$.

Step 9. Compute the forward kinematics function $\boldsymbol{T}_n^0(\boldsymbol{q})$.

This procedure is demonstrated below for the SCARA-type robot with a planar three-link arm shown in Figure 5.1 and Figure 5.2c.

5.4.4 Decoupled axes of motion

The decoupling of robot axes of motion refers to controlling one axis of motion without affecting any other axis of motion. Mechanically decoupled axes can be of interest for several reasons:

- Simplified computation of forward and inverse kinematics
- Simplified controller design
- Only one actuator needed to drive one axis of motion
- Partial robot functionality even if one axis of motion fails.

The individual operation of decoupled axes can be utilized as a safety feature: the axes can be sequentially actuated so that only one mechanism is active at a time (Stoianovici et al., 2003). While the design of robots with six mechanically decoupled axes is difficult (Schenker et al., 1995), the design of a cylindrical robot with three decoupled axes is feasible. SCARA-type, cylindrical robots with three mechanically decoupled axes are used for substrate-handling in electronics manufacturing (Aalund and Mathia, 2001; Mathia and Aalund, 2002). Example 5.4 discusses such a mechanical arm. Section 7.6 presents a measurement system for axis-decoupling tests.

Example 5.3: DH notation for a SCARA-type robot

This example establishes the DH notation for SCARA-type robots with a cylindrical coordinate frame. These robots are commonly used in electronics manufacturing, in particular in the semiconductor, flat panel display (FPD), hard disk, and solar industries, for both atmospheric and vacuum applications. Vacuum robots are located at the center of several process tools clustered in a circular arrangement around the robot (hence the term 'cluster tool'). Cylindrical robot coordinate frames are a natural choice for such configurations. Atmospheric process and metrology tools can be described in either cylindrical or Cartesian coordinate frames. In the latter case a coordinate transform maps Cartesian coordinates to the robot's cylindrical coordinate frame. For example, an equipment front end module (EFEM) is a mini-environment that handles carriers and substrates at the interface between the factory material handling system and the process equipment. EFEMs are naturally described in Cartesian coordinates. See SEMI Standard E101-1104.

The DH notation for the SCARA-type robot in Figure 5.1 and Figure 5.2c is established using the above procedure. The robot has $n=4$ joints, including one vertical axis of motion and three axes for the planar three-link arm. The third link is the end-effector. Refer to Figure 5.10 for an illustration of the DH parameters and local coordinate frames per arm link.

Step-by-step procedure:

Step 1. All joints are identified and numbered $i=1,\ldots,4$. The prismatic joint 1 provides vertical motion; joints 2, 3, and 4 are revolute joints and part of the planar arm (shoulder, elbow, and wrist).

Step 2. All links are identified and numbered. Link 0 is the robot base; link 1 is controlled by the prismatic joint 1; the 'inner' link 2 is controlled by the revolute joint 2; the 'outer' link 3 is controlled by the revolute joint 3; the end-effector (link 4) is controlled by the revolute joint 4.

Step 3. Attach coordinate frames to all links; see Figure 5.10.

Step 4. Origin O_1 is at the end of link 1.

Step 5. Origins O_2, O_3, and O_4 are at the end of links 2, 3, and 4, respectively. (Note: for demonstration purposes the vertical offset of each individual revolute joint is modeled here, although they can be conveniently combined with the prismatic joint position d_1.)

Step 6. The end-effector's x_4-axis intersects with the revolute joint axis 3 at a right angle. The z_4-axis is parallel to z_3.

Step 7. The DH parameters are listed in Table 5.1.

Step 8. The generalized coordinate vector $\boldsymbol{q} \in \mathbb{R}^4$ includes the coordinates

$$\begin{aligned} q_1 &= d_1, \text{ joint position for prismatic joint 1} \\ q_i &= \theta_i, \text{ joint angles for revolute joints with } i = 2, 3, 4. \end{aligned} \tag{5.26}$$

The homogeneous transform in Equation (5.23) also applies to revolute joints 2 to 4, with the transformation matrices

$$A_4^3(\theta_4) = \begin{bmatrix} \cos(\theta_4) & -\sin(\theta_4) & 0 & l_4 \cdot \cos(\theta_4) \\ \sin(\theta_4) & \cos(\theta_4) & 0 & l_4 \cdot \sin(\theta_4) \\ 0 & 0 & 1 & d_4 \\ 0 & 0 & 0 & 1 \end{bmatrix} \tag{5.27}$$

$$A_3^2(\theta_3) = \begin{bmatrix} \cos(\theta_3) & -\sin(\theta_3) & 0 & l_3 \cdot \cos(\theta_3) \\ \sin(\theta_3) & \cos(\theta_3) & 0 & l_3 \cdot \sin(\theta_3) \\ 0 & 0 & 1 & d_3 \\ 0 & 0 & 0 & 1 \end{bmatrix}. \tag{5.28}$$

The transform from link 2 to link 1 involves a translation and a rotation:

$$A_2^1(\theta_2) = \begin{bmatrix} \cos(\theta_2) & -\sin(\theta_2) & 0 & l_2 \cdot \cos(\theta_2) \\ \sin(\theta_2) & \cos(\theta_2) & 0 & l_2 \cdot \sin(\theta_2) \\ 0 & 0 & 1 & 0 \\ 0 & 0 & 0 & 1 \end{bmatrix}. \tag{5.29}$$

The transform from link 1 to the base link 0 is a translation only:

$$A_1^0(d_1) = \begin{bmatrix} 1 & 0 & 0 & 0 \\ 0 & 1 & 0 & 0 \\ 0 & 0 & 1 & d_1 \\ 0 & 0 & 0 & 1 \end{bmatrix}. \tag{5.30}$$

Step 9. The homogeneous transform from end-effector coordinates to the reference frame $O_0\text{-}x_0y_0z_0$ is then

$$\tilde{\boldsymbol{x}}_0 = \boldsymbol{T}_4^0(\boldsymbol{q}) \cdot \tilde{\boldsymbol{x}}_4. \tag{5.31}$$

The following notation is used: $\cos(\theta_i + \theta_j) = \mathrm{c}_{ij}$, $\sin(\theta_i + \theta_j) = \mathrm{s}_{ij}$. The overall homogeneous transformation matrix $\boldsymbol{T}$ becomes

$$\begin{aligned} \boldsymbol{T}_4^0(\boldsymbol{q}) &= \boldsymbol{A}_1^0(d_1) \cdot \boldsymbol{A}_2^1(\theta_2) \cdot \boldsymbol{A}_3^2(\theta_3) \cdot \boldsymbol{A}_4^3(\theta_4) \\ &= \begin{bmatrix} 1 & 0 & 0 & 0 \\ 0 & 1 & 0 & 0 \\ 0 & 0 & 1 & d_1 \\ 0 & 0 & 0 & 1 \end{bmatrix} \cdot \begin{bmatrix} \mathrm{c}_2 & -\mathrm{s}_2 & 0 & \mathrm{c}_2 l_2 \\ \mathrm{s}_2 & \mathrm{c}_2 & 0 & \mathrm{s}_2 l_2 \\ 0 & 0 & 1 & 0 \\ 0 & 0 & 0 & 1 \end{bmatrix} \cdot \\ &\quad \begin{bmatrix} \mathrm{c}_3 & -\mathrm{s}_3 & 0 & \mathrm{c}_3 l_3 \\ \mathrm{s}_3 & \mathrm{c}_3 & 0 & \mathrm{s}_3 l_3 \\ 0 & 0 & 1 & d_3 \\ 0 & 0 & 0 & 1 \end{bmatrix} \cdot \begin{bmatrix} \mathrm{c}_4 & -\mathrm{s}_4 & 0 & \mathrm{c}_4 l_4 \\ \mathrm{s}_4 & \mathrm{c}_4 & 0 & \mathrm{s}_4 l_4 \\ 0 & 0 & 1 & d_4 \\ 0 & 0 & 0 & 1 \end{bmatrix} \\ &= \begin{bmatrix} \mathrm{c}_{234} & -\mathrm{s}_{234} & 0 & l_2\mathrm{c}_2 + l_3\mathrm{c}_{23} + l_4\mathrm{c}_{234} \\ \mathrm{s}_{234} & \mathrm{c}_{234} & 0 & l_2\mathrm{s}_2 + l_3\mathrm{s}_{23} + l_4\mathrm{s}_{234} \\ 0 & 0 & 1 & d_1 + d_3 + d_4 \\ 0 & 0 & 0 & 1 \end{bmatrix}. \end{aligned} \tag{5.32}$$

This completes the procedure.

The following well-known trigonometric simplifications are used to obtain the solution in Equation (5.32); note the reverse signs in the second relationship:

$$\begin{aligned}\cos(\alpha)\cdot\sin(\beta)\pm\cos(\beta)\cdot\sin(\alpha)&=\sin(\alpha\pm\beta)\\ \cos(\alpha)\cdot\cos(\beta)\pm\sin(\alpha)\cdot\sin(\beta)&=\cos(\alpha\mp\beta)\end{aligned}. \tag{5.33}$$

Also note that elements (1,4) and (2,4) of $\boldsymbol{T}_4^0(\boldsymbol{q})$ are identical with rows 1 and 2 of the vector-valued function in Equation (5.18), which was obtained using straightforward geometry. The rotation angle in Equation (5.18), $\phi=\theta_1+\theta_3+\theta_3$, is implemented with the rotation matrix $\boldsymbol{R}_z$ within $\boldsymbol{T}_4^0(\boldsymbol{q})$ in Equation (5.32). For arms with only revolute joints all offsets can be combined into one offset parameter, therefore $z_0=d_1+d_3+d_4$. Table 5.2 lists the end-effector pose in the reference frame O_0-$x_0y_0z_0$ for a given set of DH parameters, computed with Equations (5.31) and (5.32). □

Table 5.1. Denavit–Hartenberg parameters for Example 5.3.

Link	l_i	α_i	d_i	θ_i
1	0	0	d_1	0
2	l_2	0	0	θ_2
3	l_3	0	d_3	θ_3
4	l_4	0	d_4	θ_4

Table 5.2. End-effector pose in the reference frame for specified DH parameters.

	DH parameters				End-effector pose			
Link	l_i	α_i	d_i	θ_i	x_0	y_0	z_0	θ_0
1	0	0	1	0	−1.0	0.0	1.3	$3\pi/2$
2	1	0	0	$\pi/2$				
3	1	0	0.2	$\pi/2$				
4	1	0	0.1	$\pi/2$				

5.4.5 Differential kinematics and kinematic singularities

The differential kinematics of a robot describes the relationship between joint velocities and the corresponding end-effector velocities, both translational and rotational. The method is based on the linearization principle, a common method for

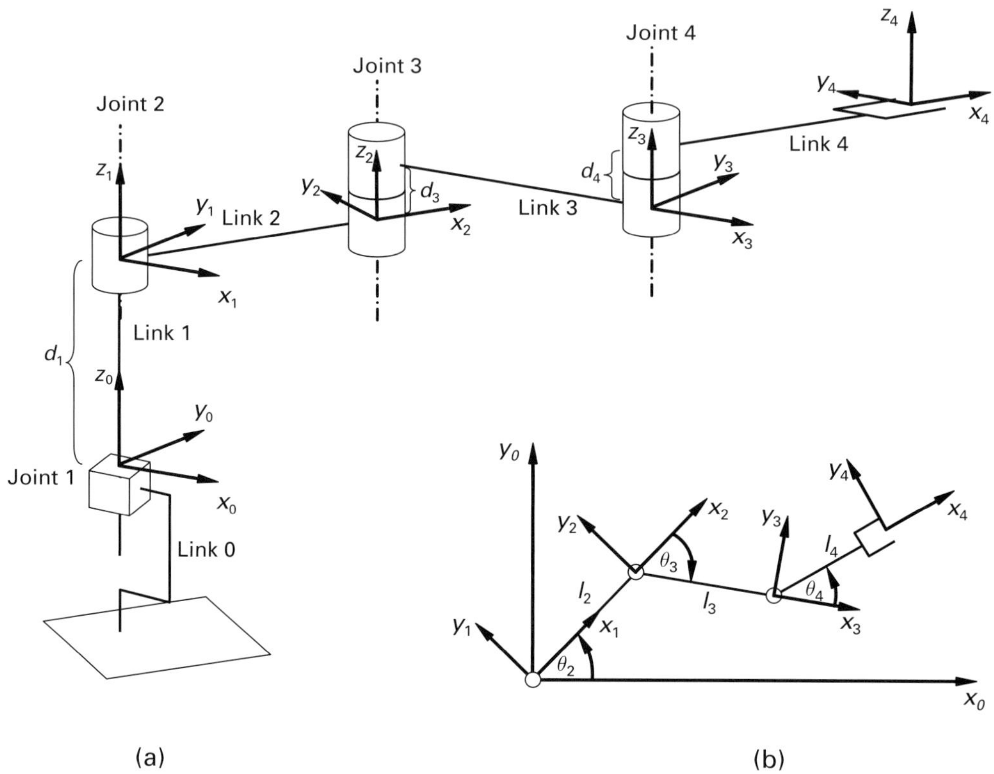

Figure 5.10 Denavit–Hartenberg notation for the robot in Example 5.3.

Example 5.4: cylindrical robot with decoupled axes

This example presents a cylindrical, SCARA-type robot with three mechanically decoupled axes of motion. Only one active motor is required to control one axis of motion, which significantly simplifies the control algorithm. The robot arm in Figure 5.11 has decoupled axes. The DH notation from Example 5.3 is used. The arm has two links of identical length $l_2 = l_3 = l$, measured between the joints' axes of rotation. The third link is the end-effector, whose length is measured from the joint to the R-position R_4, which is also the origin O_4. Pulleys 2 and 3 are concentric and firmly attached to the same shaft. All joint angles θ_i are defined counterclockwise.

Decoupled R-axis. The arm is mechanically constrained to radial motion and is controlled by θ_2. Joint angle θ_2 is driven by the actuator for joint 2. The end-effector reference position R_4 and the axis of rotation of joint 4 (and pulley 4) are always located on the R-axis (Figure 5.11). The end-effector is therefore always orientated along the R-axis. The constant end-effector orientation along the R-axis is achieved with pulley ratios of 2:1 within the inner link (pulley 1/pulley 2) and 1:2 within the outer link (pulley 3/pulley 4).

This results in the following joint angle relationships:

$$\begin{aligned}\theta_3 &= 180^\circ - 2\cdot\theta_2\\ \theta_4 &= 90^\circ - \theta_2.\end{aligned} \tag{5.34}$$

The forward kinematics from straightforward geometry is:

$$R = R_4 = 2\cdot l\cdot \sin(\theta_2) + l_4. \tag{5.35}$$

The radial end-effector position R_4 is provided as a function of the controlled joint angle θ_2 only.

The same result is obtained with DH notation, which can be verified by comparing the $\boldsymbol{T}_4^0$ matrix (in Cartesian coordinates) in Equation (5.32) with the forward kinematics (in polar coordinates) in Equation (5.35). The comparison requires a coordinate transform from Cartesian to polar coordinates, which is easily tested for the special case of motion along the y_0-axis only: the element (1,4) in $\boldsymbol{T}_4^0$ becomes $x_0 = 0$. Element (2,4) in $\boldsymbol{T}_4^0$ becomes

$$y_0 = l_2\cdot\sin(\theta_2) + l_3\cdot\sin(\theta_2+\theta_3) + l_4\cdot\sin(\theta_2+\theta_3+\theta_4). \tag{5.36}$$

Note that θ_4 in Figure 5.11 is clockwise and thus negative. With the relationships in Equation (5.34), $\theta_3 = 180^\circ - 2\theta_2$, and $\theta_4 + 2\theta_2 + (90^\circ - \theta_2) = 180^\circ$ for the triangle formed by the two arm links, the sine functions reduce to

$$\begin{aligned}\sin(\theta_2+\theta_3) &= \sin(\theta_2 + (180^\circ - 2\theta_2)) = \sin(180^\circ - \theta_2) = \sin(\theta_2)\\ \sin(\theta_2+\theta_3+\theta_4) &= \sin(180^\circ - \theta_2 + (-90^\circ + \theta_2)) = \sin(90^\circ) = 1.\end{aligned} \tag{5.37}$$

Therefore, with $l_2 = l_3 = l$:

$$\begin{aligned}y_0 &= l_2\cdot\sin(\theta_2) + l_3\cdot\sin(\theta_2) + l_4\\ &= 2\cdot l\cdot\sin(\theta_2) + l_4\end{aligned}. \tag{5.38}$$

This is identical to the result in Equation (5.35) from straightforward geometry. A similar result can be obtained along the x_0-axis.

Decoupled Θ-axis. In practice the robot's rotational Θ-axis can be implemented by rotating the entire drive train for the R-axis by the angle θ_1. Θ depends on the overall gear ratio M_Θ and the motor angle φ_Θ:

$$\Theta = M_\Theta\cdot\varphi_\Theta. \tag{5.39}$$

Decoupled Z-axis. The Z-axis is not shown in Figure 5.11. It is implemented with a prismatic joint, typically a lead screw. Z can be obtained from

$$Z = M_Z\cdot\varphi_Z. \tag{5.40}$$

M_Z is the overall gear ratio and φ_Z is the motor angle. □

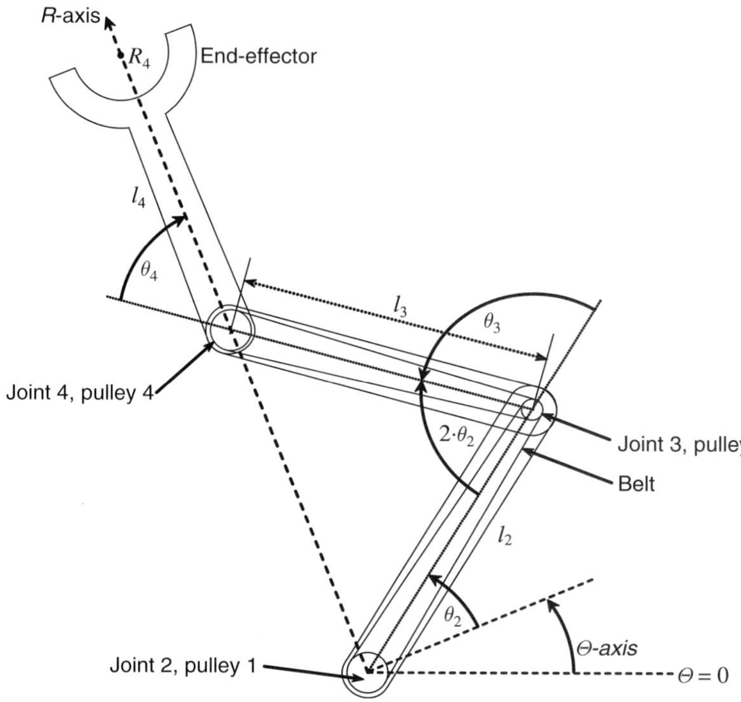

Figure 5.11 Three-link arm (including end-effector) of a cylindrical, SCARA-type robot (DH notation adopted from Example 5.3).

solving nonlinear equations. The method relies on sufficiently accurate first-order approximations of the local behavior of the nonlinear system. Here the nonlinear forward kinematics function is $\boldsymbol{x}(t) = \boldsymbol{f}(\boldsymbol{q}(t))$ in Equation (5.16) (Sontag, 1990). The linearized system is a valid representation of the original nonlinear system for small deviations from the linearization point. See, for example, (Asada and Slotine, 1986; Mathia, 1996). The velocity equation, or 'differential kinematics,' can be obtained using the time derivative of $\boldsymbol{f}(\boldsymbol{q})$, where $\boldsymbol{f}$ is the differentiable forward kinematics model in the robot's operational space:

$$\frac{d\boldsymbol{x}}{dt} = \frac{\partial \boldsymbol{f}(\boldsymbol{q})}{\partial \boldsymbol{q}} \cdot \frac{d\boldsymbol{q}}{dt} \tag{5.41}$$

$$\dot{\boldsymbol{x}} = \boldsymbol{J}(\boldsymbol{q}) \cdot \dot{\boldsymbol{q}}, \ \boldsymbol{J} \in \mathbb{R}^{m \times n}. \tag{5.42}$$

The chain rule is used on the right-hand side of Equation (5.41). A more common form is Equation (5.42). $\dot{\boldsymbol{x}}$ is the m-vector of end-effector velocities in the robot's workspace, and $\dot{\boldsymbol{q}}$ is the n-vector of generalized joint velocities in joint space. $\boldsymbol{J}(\boldsymbol{q})$ is the $m \times n$ robot Jacobian matrix (or simply Jacobian) that defines the linearized forward kinematics for a given set of joint displacements $\boldsymbol{q}$ (the arm configuration).

The Jacobian is the matrix of partial derivatives for the forward kinematics function $\boldsymbol{f}(\boldsymbol{q})$. In the most general case it is a $6 \times n$ matrix, corresponding to the general forward kinematics function in Equation (5.17):

$$J(\boldsymbol{q}) = \begin{bmatrix} \dfrac{f_1(\boldsymbol{q})}{\partial \boldsymbol{q}} \\ \vdots \\ \dfrac{f_m(\boldsymbol{q})}{\partial \boldsymbol{q}} \end{bmatrix} = \begin{bmatrix} \dfrac{df_1(\boldsymbol{q})}{dq_1} & \cdots & \dfrac{df_1(\boldsymbol{q})}{dq_n} \\ \vdots & & \vdots \\ \dfrac{df_m(\boldsymbol{q})}{dq_1} & \cdots & \dfrac{df_m(\boldsymbol{q})}{dq_n} \end{bmatrix}. \tag{5.43}$$

The Jacobian is an important tool for robot analysis and is widely used for identifying kinematic singularities and redundancies of arm configurations, solving inverse kinematics problems, analyzing forces and torques, and deriving robot arm dynamics. A wealth of analytical and numerical tools is available for the analysis of linear systems.

A kinematic singularity occurs if the Jacobian matrix has less than full rank for a given set of joint displacements $\boldsymbol{q}$, in which case its determinant is zero,

$$\det(J(\boldsymbol{q})) = 0. \tag{5.44}$$

In robotics it is important to find and understand singularities for several reasons. The following are typical at or near singularities:

- Robots lose one degree of freedom, therefore mobility is restricted.
- Small velocities in workspace require large velocities in joint space, therefore motion control can be challenging.
- Many or an infinite number of solutions to the inverse kinematics problem may exist.

Obvious kinematic singularities are the outer and inner (if applicable) workspace boundaries: the robot loses one degree of freedom and cannot extend or retract the arm beyond those limits. Boundary singularities are easy to avoid. However, depending on the arm type there can be less intuitive 'internal singularities' that can only be found using mathematical analysis.

5.4.6 Kinematic and compliance calibration

Calibration can be described as a set of operations that establish the relationship between measured quantities and the corresponding reference values (SEMI standard E89-1104). Two different kinematic calibration methods are common in industrial robotics. The first method calibrates the robot's joint displacement sensors with the robot's reference or 'home' position (Asada and Slotine, 1986). This is called 'homing' and is performed by the robot operator after system start-up or when a robot has lost its home position, for example after a power failure. Automatic homing is desirable for reducing labor and tool downtime. The second method, kinematic calibration, improves a robot's positioning accuracy, and is usually performed by the robot manufacturer to guarantee the published product specifications (Sciavicco and Siciliano, 1996). Kinematic calibration offers performance and cost advantages for a robot product.

Kinematic calibration

Forward kinematics represents an ideal, nominal robot, "the unique, theoretical model representing a robot built with zero tolerance" (RIA standard R15.07). In practice, however, robots are made of mechanical and electromechanical components whose

Example 5.5: singularities of a planar three-link arm

Consider the planar three-link robot arm illustrated in Figure 5.1. The generalized joint variable consists only of joint angles, $\boldsymbol{q} = \theta$. With the forward kinematics in Equation (5.18) the Jacobian is

$$\boldsymbol{J}(\theta) = \begin{bmatrix} \frac{df_1(\theta)}{d\theta_1} & \frac{df_1(\theta)}{d\theta_2} & \frac{df_1(\theta)}{d\theta_3} \\ \frac{df_2(\theta)}{d\theta_1} & \frac{df_2(\theta)}{d\theta_2} & \frac{df_2(\theta)}{d\theta_3} \\ \frac{df_3(\theta)}{d\theta_1} & \frac{df_3(\theta)}{d\theta_2} & \frac{df_3(\theta)}{d\theta_3} \end{bmatrix} \tag{5.45}$$

$$= \begin{bmatrix} -l_1s_1 - l_2s_{12} - l_3s_{123} & -l_2s_{12} - l_3s_{123} & -l_3s_{123} \\ l_1c_1 + l_2c_{12} + l_3c_{123} & l_2c_{12} + l_3c_{123} & l_3c_{123} \\ 1 & 1 & 1 \end{bmatrix}. \tag{5.46}$$

$\boldsymbol{J}(\theta)$ is singular if $\det(\boldsymbol{J}(\theta)) = 0$:

$$\det(\boldsymbol{J}(\boldsymbol{q})) = l_1c_1 \cdot l_2c_{12} - l_1s_1 \cdot l_2s_{12} = 0. \tag{5.47}$$

The robot's singularities are therefore the inner workspace boundary with $\theta_2 = 0$ and the outer workspace boundary with $\theta_2 = \pi$. The joint angles θ_1 and θ_3 are arbitrary. These are the boundaries of the dexterous workspace where the end-effector has all degrees of freedom. The Jacobian here is reduced to a 3×3 matrix, from the general 6×3 matrix (Sciavicco and Siciliano, 1996):

- For cylindrical robots the Z-axis is decoupled from the planar arm and is therefore not considered.
- End-effector rotation is mechanically possible only about the workspace z-axis, not about the x and y axes.

The reduced Jacobian facilitates the analysis and implementation of control algorithms. In the general case the $6{\times}n$ Jacobian corresponding to the forward kinematics in Equation (5.17) comprises the Jacobian $\boldsymbol{J}_\mathrm{L} \in \mathbb{R}^{3\times3}$ for linear (translational) end-effector motion, and $\boldsymbol{J}_\mathrm{R} \in \mathbb{R}^{3\times3}$ for rotary (angular) motion:

$$\begin{bmatrix} \dot{\boldsymbol{p}} \\ \dot{\phi} \end{bmatrix} = \begin{bmatrix} \boldsymbol{J}_\mathrm{L}(\theta) \\ \boldsymbol{J}_\mathrm{R}(\theta) \end{bmatrix} \dot{\theta}$$

$$\begin{bmatrix} \dot{x} \\ \dot{y} \\ \dot{z} \\ \dot{\varphi} \\ \dot{\vartheta} \\ \dot{\psi} \end{bmatrix} = \begin{bmatrix} -l_1s_1 - l_2s_{12} - l_3s_{123} & -l_2s_{12} - l_3s_{123} & -l_3s_{123} \\ l_1c_1 + l_2c_{12} + l_3c_{123} & l_2c_{12} + l_3c_{123} & l_3c_{123} \\ 0 & 0 & 0 \\ 0 & 0 & 0 \\ 0 & 0 & 0 \\ 1 & 1 & 1 \end{bmatrix} \begin{bmatrix} \dot{\theta}_1 \\ \dot{\theta}_2 \\ \dot{\theta}_3 \end{bmatrix} \tag{5.48}$$

Only the non-zero rows contribute to robot motion. The zero rows indicate that the z, ϑ, ψ velocities are zero, regardless of the joint angle velocities. Zero rows are usually omitted. □

imperfections accumulate and affect the end-effector position in workspace. Imprecisions in the manufacturing process also contribute to deviations from the ideal robot. Typical robot components with tolerances include arm links, bearings, lead screws, gears and transmissions (belts and pulleys, rack and pinion), electric motors, and encoders. Kinematic calibration as a standard procedure for each robot prior to delivery and installation can have economic advantages: determining accurate kinematic parameters may be more cost-efficient than manufacturing high-precision parts. The estimated parameters must permanently reside with each individual robot, for example stored in flash memory inside the robot base. The parameters can be queried during start-up.

Kinematic calibration is typically restricted to the static analysis of the end-effector position as a result of variations in factors such as

- link lengths, joint axis orientation, robot base location (geometric effects)
- gear eccentricity, backlash, joint compliance due to gravity (non-geometric effects).

The concept is to compensate an individual robot for its physical imperfections by measuring a series of end-effector positions and estimating the Denavit–Hartenberg parameters based on the measured position errors. The estimates allow improvement and 'customizing' of the robot's forward kinematics model. For example, in one study a PUMA 560 robot had an original end-effector accuracy of about 0.01 m, which could be improved to 0.0003 m (Hollerbach, 1989). Kinematic calibration involves the numerical inversion of the robot's forward kinematics and a precision position measurement system.

The following step-by-step process addresses geometric effects (Hollerbach, 1989):

- Step 1. Establish a forward kinematics model $\boldsymbol{f}$. The end-effector position $\boldsymbol{x}$ is a function of the four Denavit–Hartenberg (kinematic) parameters, that is, the vectors $\boldsymbol{l} = [l_1, \ldots, l_n]^T$, $\alpha = [\alpha_1, \ldots, \alpha_n]^T$, $\boldsymbol{d} = [d_1, \ldots, d_n]^T$, $\theta = [\theta_1, \ldots, \theta_n]^T$. Then:

$$\boldsymbol{x} = \boldsymbol{f}(\boldsymbol{a}, \alpha, \boldsymbol{d}, \theta). \tag{5.49}$$

- Step 2. Linearize the forward kinematics model:

$$\Delta \boldsymbol{x} = \frac{\partial \boldsymbol{f}(\boldsymbol{l})}{\partial \boldsymbol{l}} \cdot \Delta \boldsymbol{a} + \frac{\partial \boldsymbol{f}(\alpha)}{\partial \alpha} \cdot \Delta \alpha + \frac{\partial \boldsymbol{f}(\boldsymbol{d})}{\partial \boldsymbol{d}} \cdot \Delta \boldsymbol{d} + \frac{\partial \boldsymbol{f}(\theta)}{\partial \theta} \cdot \Delta \theta \tag{5.50}$$

$$= \boldsymbol{J}_l(\boldsymbol{l}) \cdot \Delta \boldsymbol{l} + \boldsymbol{J}_\alpha(\alpha) \cdot \Delta \alpha + \boldsymbol{J}_d(\boldsymbol{d}) \cdot \Delta \boldsymbol{d} + \boldsymbol{J}_\theta(\theta) \cdot \Delta \theta. \tag{5.51}$$

  The partial derivatives are the Jacobians in Equation (5.45) with respect to one of the four Denavit–Hartenberg parameter vectors. The Jacobians are computed using the given kinematic parameters. $\Delta \boldsymbol{x}$ is the measured end-effector pose error for a given iteration step, and $\Delta \boldsymbol{l}$, $\Delta \alpha$, $\Delta \boldsymbol{d}$, $\Delta \theta$ are the resulting adjustments to the Denavit–Hartenberg parameters to be used for the next iteration.
- Step 3. Group the kinematic parameter vectors and partial derivatives (Jacobians) into one single, linear system, the *kinematic calibration equation*:

$$\Delta \boldsymbol{x} = \begin{bmatrix} \boldsymbol{J}_l(\boldsymbol{l}) & \boldsymbol{J}_\alpha(\alpha) & \boldsymbol{J}_d(\boldsymbol{d}) & \boldsymbol{J}_\theta(\theta) \end{bmatrix} \begin{bmatrix} \Delta \boldsymbol{l} \\ \Delta \alpha \\ \Delta \boldsymbol{d} \\ \Delta \theta \end{bmatrix} \\ = \Phi(\psi) \cdot \Delta \psi \tag{5.52}$$

This system of linearized equations is similar to the differential kinematics in Equation (5.41). It is underdetermined because the number of kinematic parameters, $4n$, exceeds the number of equations, m. It is advisable to take M measurements such that $M \cdot m \gg 4 \cdot n$ in order to avoid an ill-conditioned matrix (Sciavicco and Siciliano, 1996). The results are M variations of Equation (5.52), which are arranged as

$$\Delta \tilde{\boldsymbol{x}} = \begin{bmatrix} \Delta \boldsymbol{x}_1 \\ \vdots \\ \Delta \boldsymbol{x}_M \end{bmatrix} = \begin{bmatrix} \Delta \Phi_1(\psi) \\ \vdots \\ \Delta \Phi_M(\psi) \end{bmatrix} = \tilde{\Phi}(\psi) \cdot \Delta \psi. \tag{5.53}$$

- Step 4. Invert the kinematic calibration equation for adjusting the kinematic parameters. A generalized inverse is needed because Equation (5.52) is underdetermined. The left Moore–Penrose pseudo-inverse Φ^+ is used here, which computes a 'best fit' solution in terms of least squares (Penrose, 1955):

$$\Delta \psi = \left(\left(\tilde{\Phi}^T(\psi) \cdot \tilde{\Phi}(\psi) \right)^{-1} \cdot \tilde{\Phi}^T(\psi) \right) \cdot \Delta \boldsymbol{x} \\ = \tilde{\Phi}^+(\psi) \cdot \Delta \boldsymbol{x}. \tag{5.54}$$

- Step 5. Numerically solve the nonlinear Equation (5.49), using the linearized Equation (5.54) and an iterative process based on Newton's method. With $\Delta \psi(n) = \psi(n+1) - \psi(n)$ at the n-th iteration, the algorithm is

$$\psi(n+1) = \psi(n) + \Delta \psi(n) \\ = \psi(n) + \tilde{\Phi}^+(\psi) \cdot \Delta \boldsymbol{x}. \tag{5.55}$$

The algorithm is illustrated in Figure 5.12. Note that the commanded and actual final end-effector positions, $\boldsymbol{x}_c$ and $\boldsymbol{x}_{final}$, are constant, thus the position error $\Delta \boldsymbol{x}$ is determined only at the first iteration ($n = 1$) and then remains constant. This is indicated with the dashed, outer loop in Figure 5.12. The initial values for $\psi(n = 0)$ are the nominal values from the forward kinematics model. The numerical process is terminated when $\Delta \psi$ converges within a specified limit.

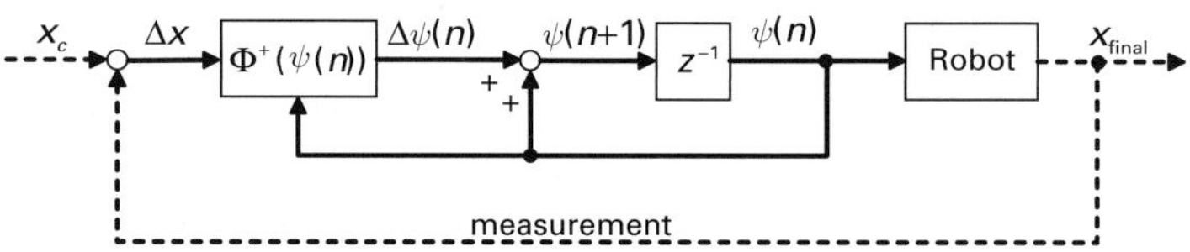

Figure 5.12 Iterative kinematic calibration process at the final end-effector position.

Newton's method, and thus the above algorithm, is one of many numerical methods for solving systems of nonlinear equations. It converges to the 'best' solution if the initial point $\psi(n = 0)$ is sufficiently close to the solution, and if the function $\boldsymbol{f}(\boldsymbol{l}, \boldsymbol{\alpha}, \boldsymbol{d}, \boldsymbol{\theta})$ in Equation (5.49) is continuously differentiable. Here the nominal parameters are used as the initial condition. The differentiability condition excludes possible singularities of $\boldsymbol{f}$, which would cause numerical instability problems if approached by $\psi(n)$. This can occur if adjacent joint axes are (nearly) parallel, as is the case for the SCARA-type robots discussed here. A modification of the Denavit–Hartenberg parameters to avoid this has been proposed (Hayati and Mirmirani, 1985).

Compliance calibration

The compliance of a robot manipulator refers to the inherent elasticity in the drive trains, links, and bearings. Given a sufficient load, a manipulator deforms, and as a result the end-effector position deviates from the commanded position. The typical measurements of joint displacements for robot control cannot directly account for the deformation of links and the compliance or backlash in drive train components. The actual end-effector position under load may therefore be unknown. The deformation can be accounted for with a suitable compliance model and an advanced robot control algorithm. Such a strategy involves the development of analytical tools and experimental procedures to accurately establish a robot compliance model. In (Hudgens, 1992; Hudgens et al., 2000) a lumped model approximation is used in conjunction with a quasistatic analysis to describe the gross deflection of a general serial manipulator that includes both joint and link compliance sources. The quasistatic assumption is valid if the frequency content of the loads is well below the fundamental frequency of the arm. A parameter estimation technique determines parameters of the manipulator compliance model, which can be applied in real-time deflection compensation schemes. For example, some commercially available substrate-handling robots use the active leveling of the end-effector to compensate for compliance issues.

5.5 Inverse kinematics

The RIA defines the inverse kinematics model of a robot as "the mathematical relationship which determines the principal joint displacements based on the values of the robot tool coordinate system TCS" (RIA standard R15.07). The Cartesian coordinate system is a common TCS, as is the cylindrical coordinate system of SCARA-type substrate-handling robots.

Given the forward kinematics model $\boldsymbol{f}: \mathbb{R}^n \mapsto \mathbb{R}^m$ in Equation (5.16), which maps the generalized joint variables $\boldsymbol{q}$ to the end-effector position and orientation $\boldsymbol{x}$, the inverse kinematics model $\boldsymbol{f}^{-1}: \mathbb{R}^m \mapsto \mathbb{R}^n$ determines the required $\boldsymbol{q}$ to achieve a desired $\boldsymbol{x}$:

$$\boldsymbol{q} = \boldsymbol{f}^{-1}(\boldsymbol{x}), \ \boldsymbol{q} \in \mathbb{R}^n, \boldsymbol{x} \in \mathbb{R}^m. \tag{5.56}$$

The inverse kinematics problem is to find the function $\boldsymbol{f}^{-1}$. Analytical solutions may not exist, due to nonlinearities and redundancies of $\boldsymbol{f}$. Manipulator redundancy is given if more than one set of joint displacements, $\boldsymbol{q}$, achieves the desired $\boldsymbol{x}$. This is the case if there

Table 5.3. Maximum number of solutions for six-DOF robots, after (Tolani et al., 2000).

No. of joint variables, n	No. of solutions (upper bound)	Comment
< 6	0	Over-constrained system
> 6	∞	Under-constrained system
6R, 5R1P	16	
4R2P	8	SCARA-type
3R3P	2	SCARA-type

are more unknown joint displacements than workspace dimensions, or $n > m$. The robot end-effector position and orientation in a three-dimensional workspace provides six equations (constraints). If redundant solutions exist a common strategy is to select the solution that is optimal, in some sense, among all possible solutions. Table 5.3 lists upper bounds on the number of solutions for robots with six degrees of freedom. R denotes a revolute joint and P a prismatic joint; for example, '5R1P' is a robot with five revolute joints and one prismatic joint.

Numerical methods can be used to overcome challenging nonlinearities in forward kinematics. In practice this computation is performed by robot controllers in real time. The algorithm in Figure 5.12 is one example of a numerical solution based on Newton's method. Many solutions to the inverse kinematics problem have been proposed in the literature. The interested reader is referred, for example, to (Asada and Slotine, 1986; Sciavicco and Siciliano, 1996; Tolani et al., 2000).

A solution to the inverse kinematics problem does exist if three consecutive revolute joints axes are parallel (Sciavicco and Siciliano, 1996). This is true for SCARA-type robots with three-link planar arms, including end-effector, as discussed in the following example.

5.6 Commercial substrate-handling robots

Many commercial SCARA-type substrate-handling robots are available for electronics manufacturing. The following examples demonstrate the variety of robot designs based on the SCARA-type kinematic structure.

5.6.1 Robots arms with constrained arm motion

A single-arm robot with constrained arm motion is discussed in Example 5.4 and Example 5.6. This concept has been used for planar, four-link arms (Figure 5.14a). The motivation for this design is the longer arm reach, and therefore a larger workspace. Dual-arm robots leverage single-arm robot designs, with the addition of an arm and drive train. The motivation for this design is to improve substrate throughput when the same substrate carrier is loaded and unloaded. The robot in Figure 5.2d has two identical and symmetric arms, optionally with the mechanical

Example 5.6: inverse kinematics of a mechanically constrained three-link arm

Consider the planar three-link robot arm and a cylindrical workspace in Figure 5.11. The end-effector is the third link. The DH notation is adopted from Example 5.3. The arm has two solutions to the inverse kinematics problem and is therefore redundant (Figure 5.13). This is also noted in Table 5.3 for 3R3P robots. The forward kinematics model in Equation (5.35) can be inverted analytically:

$$\theta_2 = \sin^{-1}\left(\frac{R_4 - l_4}{2l}\right), \text{ with } -2l + l_4 \leq R_4 \leq 2l + l_4. \tag{5.57}$$

The radial end-effector position R_4 is controlled with the joint angle θ_2. The inverse sine function is defined for $-90^\circ \leq \theta_2 \leq +90^\circ$, that is, from full arm retraction to full extension. Hard stops in shoulder and elbow can be used to restrict θ_2 to the admissible range. □

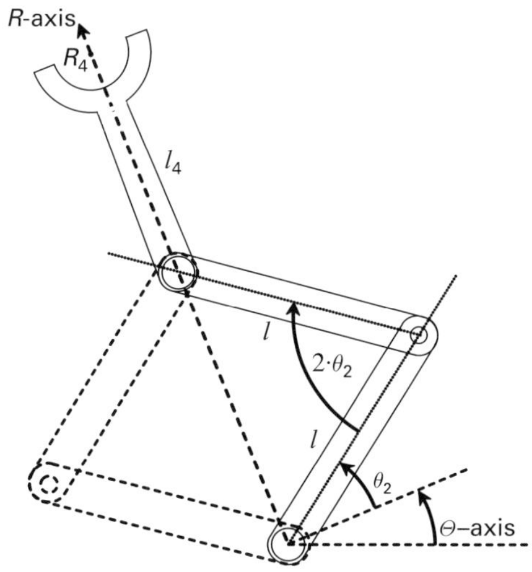

Figure 5.13 Redundancy: two solutions to the inverse kinematics problem (Denavit–Hartenberg notation adopted from Example 5.3.).

constraints discussed above. The product commonality in the design of both arms reduces manufacturing and maintenance costs. Both arms share one rotational Θ-axis and one vertical Z-axis in a cylindrical coordinate frame: they move along the same R-axis at almost the same vertical position. Most commercial dual-arm robots allow the independent extension and retraction of both arms, although a few models utilize one drive train for both arms and mechanically force opposing R-motions in each arm.

The advantages of constrained arm motion include the simple and cost-efficient design, a reduced number of motion axes and drive trains, and consequently a simpler robot controller. A disadvantage is the lack of flexibility for a wider range of applications. Some single-arm and dual-arm robots use parallelogram mechanisms for their multi-link arms (Figure 5.14c). Motivations include increased stiffness for higher payloads and the avoidance of belts in the drive trains.

5.6.2 Robot arms with independent, fully controllable arm links

The above kinematic robot structures for single-arm and dual-arm robots are available in variations with increasing levels of complexity, up to two fully independent arms with three independently controlled joints (shoulder, elbow, and wrist). These robots are designed for manufacturing tools that are not arranged in circular cluster configurations, but in rectangular configurations describable in Cartesian coordinates. One example is mini-environments such as equipment front end modules (EFEMs) with two or three aligned substrate carriers. Such systems can be loaded and unloaded with a four-axis single arm robot (Figure 5.14d). Fully controllable single-arm robots are used for both atmospheric and vacuum applications in semiconductor, flat panel display (FPD), and data storage manufacturing. FPD-handling robots have larger workspaces, stronger links,

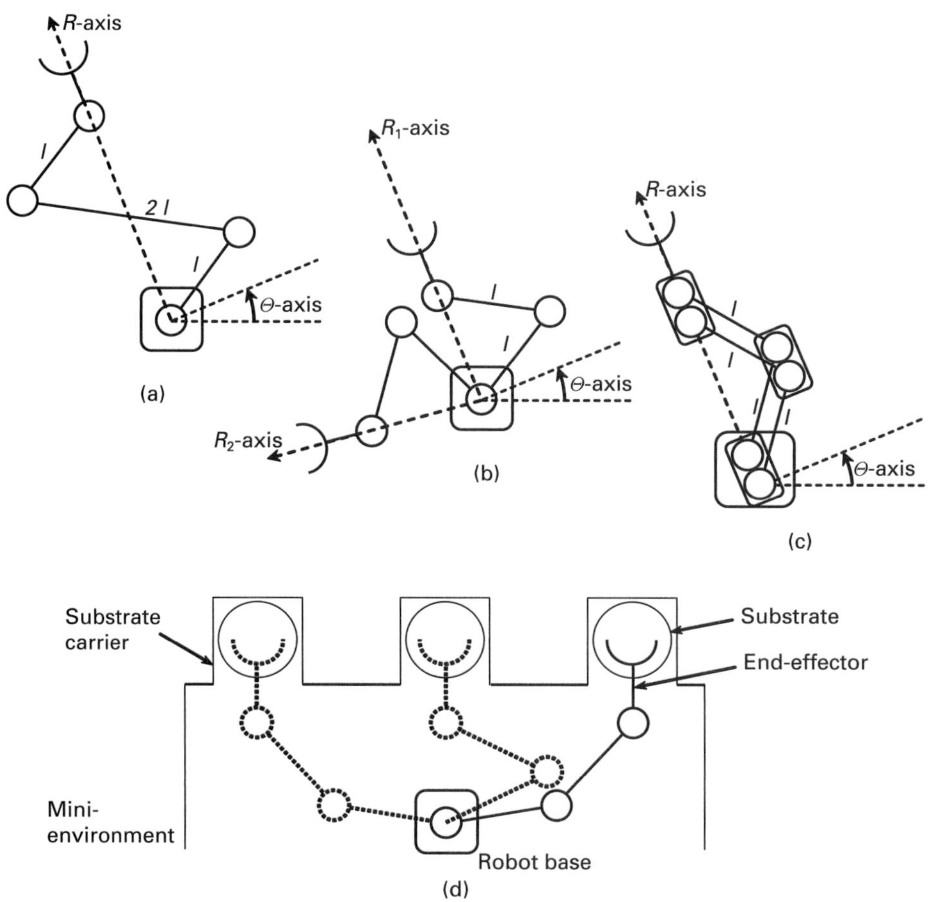

Figure 5.14 Skeleton structures of robot arms. See text for details.

heavy-duty revolute and prismatic joints, and more powerful motors that enable the transfer of increased payloads. An alternative to fully controllable arms is a constrained arm and a linear track that provides an additional axis of motion for moving the entire robot.

5.6.3 Closed kinematic chains

Some vacuum robot models utilize one or two closed kinematic chains of multi-link (sometimes called 'frog-leg') mechanisms (Figure 5.2d). These arms do not require drive belts, but they are more complex than is generally perceived: the design includes twice as many links and bearings and uses belts in the joints to constrain the number of degrees of freedom from four to two (rotation and extension). Popular commercial vacuum robots utilize one- or two-sided end-effectors, or even two 'frog-leg' arms, for increased-throughput applications. The end-effectors are extended either in the same or in opposite directions.

5.7 Applicable and related standards

Several industry standards and guidelines apply directly or indirectly to the robots discussed in this chapter. The following list provides a selection. ISO standards are published by the International Organization for Standardization (ISO). RIA standards are published by the Robotic Industries Association (RIA). SEMI standards are published by Semiconductor Equipment and Materials International (SEMI). Contact information for these organizations is listed in Appendix B.

ISO 8373:1996, Manipulating industrial robots. Vocabulary.

ISO 9283:1998, Manipulating industrial robots. Performance criteria and related test methods.

ISO 9409–1:1997, Manipulating industrial robots. Mechanical interfaces. Plates (form A).

ISO 9409–2:1997, Manipulating industrial robots. Mechanical interfaces. Shafts (form A).

ISO 9787:1999, Manipulating industrial robots. Coordinate systems and motion nomenclatures.

ISO 9946:1999, Manipulating industrial robots. Presentation of characteristics.

ISO 11593:1998, Manipulating industrial robots. Automatic end-effector systems. Vocabulary and presentation of characteristics.

ISO 14539:2001, Manipulating industrial robots. Object handling with grasp-type grippers. Vocabulary and presentation of characteristics.

RIA standard R15.07, Technical report on industrial robots – simulation/offline programming, Part 1: Terms, notations and data requirements for robot mode.

SEMI E2:93, Specifications for Quartz and High Temperature Wafer Carriers.

SEMI E89–1104, Guide for Measurement System Analysis.

SEMI E101, Guide for EFEM Functional Structure Model.

6 Dynamics and control

This chapter presents a generic dynamic model for industrial robot manipulators and applies it to SCARA-type substrate-handling robots used in electronics manufacturing. Further details can be found in the literature, for example in (Asada and Slotine, 1986; Spong and Vidyasagar, 1989; Sciavicco and Siciliano, 1996; Craig, 1989), and others. A decentralized, networked controller architecture is presented that offers the flexibility, modularity, and scalability needed to achieve the desired long-term cost efficiency of robot product families.

6.1 Manipulator dynamics

6.1.1 Equation of motion

The stability analysis of robot manipulators and the controller design process require a dynamical model in joint space where the robot control is executed. For example, a common trajectory tracking problem is to find a control law for the motor torques or forces as a function of time, $\tau(t)$, that achieves the trajectory of joint displacements $\boldsymbol{q}(t)$ that minimize the end-effector tracking error $\boldsymbol{e}(t) = \boldsymbol{x}_d(t) - \boldsymbol{x}(t)$ between the desired and actual trajectories in workspace. The model is obtained by combining the manipulator dynamics with the kinematics model.

A well-known dynamical model defined in joint space is

$$\boldsymbol{H}(\boldsymbol{q}) \cdot \ddot{\boldsymbol{q}} + \boldsymbol{C}(\boldsymbol{q}, \dot{\boldsymbol{q}}) \cdot \dot{\boldsymbol{q}} + \boldsymbol{G}(\boldsymbol{q}) = \boldsymbol{Q}, \tag{6.1}$$

with $\dot{\boldsymbol{q}} = \dfrac{d\boldsymbol{q}}{dt}$, $\ddot{\boldsymbol{q}} = \dfrac{d^2\boldsymbol{q}}{dt^2}$. Equation (6.1) represents a system of nonlinear differential equation with time t as the independent variable. The manipulator dynamics are a function of $\boldsymbol{q}$ and depend on the arm configuration. The left-hand side includes gravitional and inertial terms, but no external forces resulting from contact with the environment. The following notation is used. As always, n is the number of manipulator joints (degrees of freedom):

$\boldsymbol{H} \in \mathbb{R}^{n \times n}$: joint space inertia matrix (symmetric, positive definite)
$\boldsymbol{C} \in \mathbb{R}^{n \times n}$: matrix for computing the centripetal and Coriolis torques
$\boldsymbol{G} \in \mathbb{R}^{n \times 1}$: vector of gravitational forces
$\boldsymbol{q} \in \mathbb{R}^{n \times 1}$: vector of generalized joint displacements
$\boldsymbol{Q} \in \mathbb{R}^{n \times 1}$: vector of generalized actuator forces.

The generalized forces, $\boldsymbol{Q}$, are motor torques and forces applied to the manipulator joints, as well as any external forces acting on the end-effector (excluding gravity and inertial forces). The product $\boldsymbol{C}(\boldsymbol{q},\dot{\boldsymbol{q}})\cdot\dot{\boldsymbol{q}}$ is the $(n\times 1)$ vector of centripetal and Coriolis torques.

6.1.2 Lagrange's method

The parameters that describe the robot manipulator dynamics in Equation (6.1) can be derived using the well-known Lagrange or Euler methods. Lagrange's method is outlined here. For conservative mechanical systems, including robot arm linkages, the Lagrangian is defined as

$$L(\boldsymbol{q},\dot{\boldsymbol{q}}) = T(\boldsymbol{q},\dot{\boldsymbol{q}}) - V(\boldsymbol{q}). \tag{6.2}$$

$T(\dot{\boldsymbol{q}})$ is the kinetic energy and $V(\boldsymbol{q})$ is the potential energy stored in the arm linkage and motors. Since energy is additive the total energy comprises the contributions from the individual links and their associated joints and motors. However, the following discussion does not distinguish between the energy contributions from links and motors. The kinetic energy stored in the i-th arm link is a result of translational and rotational motion:

$$T_i = \frac{1}{2}m_i\boldsymbol{v}_i^T\boldsymbol{v}_i + \frac{1}{2}\omega_i^T\boldsymbol{I}_i\omega_i. \tag{6.3}$$

m_i is the mass of the i-th arm link, concentrated at CG_i, the i-th link's center of gravity. $\boldsymbol{v}_i$ is the vector of linear velocities, ω_i is the vector of angular velocities, and $\boldsymbol{I}_i$ is the inertia matrix. In the one-dimensional case Equation (6.3) is the well-known expression $T = 0.5mv^2 + 0.5I\omega^2$. Here the energy is required as a function of $\boldsymbol{q}$, which is obtained with the differential kinematics equation (5.42):

$$\boldsymbol{v}_i = \boldsymbol{J}_{L,i}\dot{\boldsymbol{q}},\ \ \boldsymbol{J}_{L,i} = [\boldsymbol{j}_{L,1}\cdots\boldsymbol{j}_{L,i}\ \boldsymbol{0}\ldots\boldsymbol{0}]. \tag{6.4}$$

$$\omega_i = \boldsymbol{J}_{R,i}\dot{\boldsymbol{q}},\ \ \boldsymbol{J}_{R,i} = [\boldsymbol{j}_{R,1}\cdots\boldsymbol{j}_{R,i}\ \boldsymbol{0}\ldots\boldsymbol{0}]. \tag{6.5}$$

$\boldsymbol{J}_{L,i}$ is the Jacobian matrix for linear (translational) motion of the i-th link, and $\boldsymbol{v}_i = \dot{\boldsymbol{x}}_i$ is the translational velocity. $\boldsymbol{J}_{R,i}$ is the Jacobian matrix for rotational motion. $\boldsymbol{j}_{L,i}$ and $\boldsymbol{j}_{R,i}$ are the i-th column vectors of $\boldsymbol{J}_{L,i}$ and $\boldsymbol{J}_{R,i}$, respectively. The motion of the i-th link depends only on joints 1,…, i, so the columns $i+1$, …, n are zero. The total $T(\dot{\boldsymbol{q}})$ for the entire arm linkage is given by

$$T(\dot{\boldsymbol{q}}) = \sum_{i=1}^{n} T_i(\dot{\boldsymbol{q}}) = \frac{1}{2}\sum_{i=1}^{n}(m_i\dot{\boldsymbol{q}}^T\boldsymbol{J}_{L,i}^T\boldsymbol{J}_{L,i}\dot{\boldsymbol{q}} + \dot{\boldsymbol{q}}^T\boldsymbol{J}_{R,i}^T\boldsymbol{I}_i\boldsymbol{J}_{R,i}\dot{\boldsymbol{q}}). \tag{6.6}$$

The vectors $\boldsymbol{j}_{L,1},\ \cdots, \boldsymbol{j}_{L,i-1}$ in $\boldsymbol{J}_{L,i}$ are identical to the vectors of the manipulator Jacobian in Equation (5.43). However, in $\boldsymbol{j}_{L,i}$ the link length l_i is replaced by the distance l_{mi} between the first joint and the centre of mass of the i-th link. $\boldsymbol{I}_i = \boldsymbol{R}_i\boldsymbol{I}_{i,i}\boldsymbol{R}_i^T$ is the rotational inertia tensor matrix of the i-th link in base coordinates.

The sum of the potential energies stored in each arm link is

$$V(\boldsymbol{q}) = \sum_{i=1}^{n} V_i(\boldsymbol{q}) = \sum_{i=1}^{n} m_i \boldsymbol{g}^T \boldsymbol{r}_i(\boldsymbol{q}). \tag{6.7}$$

$\boldsymbol{g} \in \mathbb{R}^{3\times 1}$ is the gravity vector in the robot base frame, here $\boldsymbol{g} = [0 \quad 0 \quad -g]^T$ since the z-axis of the base frame is vertical. $\boldsymbol{r}_i(\boldsymbol{q})$ is the configuration-dependent vector to CG_i in the base frame. From Equation (6.7) it follows that $V(\boldsymbol{q})$ depends on the joint variables $\boldsymbol{q}$, but not on the joint velocities $\dot{\boldsymbol{q}}$.

The Jacobians for prismatic and revolute joints can be derived using the following rule: if the i-th joint is prismatic, the i-th column of $\boldsymbol{J}_{R,i}$ is zero with $\boldsymbol{j}_{R,i} = [0\ 0\ 0]^T$. If the i-th joint is a revolute joint then $\boldsymbol{j}_{R,i} = [0\ 0\ 1]^T$. For example, a robot with one prismatic and two revolute joints has the angular velocity Jacobian

$$\boldsymbol{J}_{A,i} = \begin{bmatrix} 0 & 0 & 0 \\ 0 & 0 & 0 \\ 0 & 1 & 1 \end{bmatrix}. \tag{6.8}$$

Once T and V have been established Equation (6.1) can be derived. The Lagrangian $L(\boldsymbol{q}, \dot{\boldsymbol{q}})$ can be minimized over a given time period, which results in the generic equation of motion

$$\frac{d}{dt}\left(\frac{\partial L(\boldsymbol{q}, \dot{\boldsymbol{q}})}{\partial \dot{\boldsymbol{q}}}\right) - \frac{\partial L(\boldsymbol{q}, \dot{\boldsymbol{q}})}{\partial \dot{\boldsymbol{q}}} = \boldsymbol{Q}. \tag{6.9}$$

The parameters $\boldsymbol{H}$, $\boldsymbol{C}$, and $\boldsymbol{G}$ are defined below.

$\boldsymbol{H}$ matrix. The symmetric and positive definite $\boldsymbol{H}$ matrix is the configuration-dependent term in the kinetic energy equation (6.6). It is defined as

$$T(\dot{\boldsymbol{q}}) = \frac{1}{2}\dot{\boldsymbol{q}}^T \sum_{i=1}^{n} (m_i \boldsymbol{J}_{L,i}^T \boldsymbol{J}_{L,i} + \boldsymbol{J}_{R,i}^T \boldsymbol{I}_i \boldsymbol{J}_{R,i})\dot{\boldsymbol{q}} = \frac{1}{2}\dot{\boldsymbol{q}}^T \boldsymbol{H}\dot{\boldsymbol{q}}. \tag{6.10}$$

$\boldsymbol{C}$ matrix. The $\boldsymbol{C}$ matrix is also derived from the kinetic energy stored in the arm links. Several solutions for $\boldsymbol{C}$ may exist. The particular solution in Equation (6.11) is often used:

$$\begin{aligned} \boldsymbol{C} &= [c_{ij}],\ i = 1, \ldots, n,\ j = 1, \ldots, n \\ c_{ij} &= \sum_{k=1}^{n} c_{ij,k}\dot{q}_k = \sum_{k=1}^{n} \frac{1}{2}\left(\frac{\partial h_{ij}}{\partial q_k} + \frac{\partial h_{ik}}{\partial q_j} - \frac{\partial h_{jk}}{\partial q_i}\right)\dot{q}_k \\ c_{ij,k} &= c_{ik,j} \quad \text{(symmetry)}. \end{aligned} \tag{6.11}$$

The elements of $\boldsymbol{C}$ include partial derivatives of the inertia matrix $\boldsymbol{H}$ with respect to all joint variables. See Example 6.1.

G vector. The gravity force vector $\boldsymbol{G}$ is derived from the potential energy in Equation (6.7):

$$\boldsymbol{G} = \begin{bmatrix} G_1 \\ \vdots \\ G_n \end{bmatrix} \tag{6.12}$$

$$\begin{aligned} G_i &= \frac{\partial V}{\partial q_i} \\ &= \frac{\partial}{\partial q_i}\left(\sum_{i=1}^{n} m_i \boldsymbol{g}^T r_i(\boldsymbol{q})\right) \\ &= \sum_{i=1}^{n} m_i \boldsymbol{g}^T \frac{\partial \boldsymbol{r}_i(\boldsymbol{g})}{\partial q_i} \\ &= \sum_{i=1}^{n} m_i \boldsymbol{g}^T \boldsymbol{j}_{L,i}\ , i = 1, \ldots, \ n. \end{aligned}$$

The generalized forces $\boldsymbol{Q} \in \mathbb{R}^{n\times 1}$ are (Asada and Slotine, 1986):

$$\boldsymbol{Q} = \tau + \boldsymbol{J}^T \boldsymbol{F}_{ext}, \tag{6.13}$$

where $\tau = [\tau_1 \quad \ldots \quad \tau_n]^T$ are the actuator forces or torques exerted at the individual joints. $\boldsymbol{J}$ is the minimum form manipulator Jacobian in Equation (5.46). $\boldsymbol{F}_{ext}$ is the vector of external forces acting on the end-effector, the manipulator endpoint.

Robot control requires a solution to Equation (6.1) in the form $\boldsymbol{q} = \boldsymbol{f}(\dot{\boldsymbol{q}}, \ddot{\boldsymbol{q}})$. If analytical solutions do not exist, numerical methods are used. This involves three steps:

- Specifying the parameters $\boldsymbol{H}$, $\boldsymbol{C}$, and $\boldsymbol{G}$ in Equation (6.1) with the values for the robot in question, including the link dimensions and inertias
- Numerically solving the forward dynamics problem in robotics, that is, computing the joint accelerations $\ddot{\boldsymbol{q}}$
- Computing the joint velocities $\dot{\boldsymbol{q}}$ and position $\boldsymbol{q}$ using numerical integration.

Several algorithms are available for these steps. For example, the composite-rigid-body algorithm calculates the joint space inertia matrix $\boldsymbol{H}$. The articulated-body algorithm can be used to solve the forward dynamics problem. This algorithm can also be combined with the recursive Newton–Euler algorithm to calculate the matrix $\boldsymbol{C}$, and with a linear equation solver to compute the accelerations $\ddot{\boldsymbol{q}}$, which solves the forward dynamics problem. Variations of these algorithms, as well as other solutions, are also available. The interested reader is referred to the literature, for example (Ascher et al., 1997; Featherstone, 2007).

Example 6.1: dynamical model for a SCARA-type robot

This example establishes the dynamic parameters $\boldsymbol{H}$, $\boldsymbol{C}$, and $\boldsymbol{G}$ in Equation (6.1) for the SCARA-type robot in Figure 6.1 using Lagrange's method. The robot arm is mechanically constrained such that link 3, the end-effector, is always aligned with the R-axis. The arm parameters are shown in Figure 6.1. The Denavit–Hartenberg notation is adopted from Example 5.3.

The joint space inertia matrix $\boldsymbol{H}$ for the arm is the sum of the inertia matrices associated with the $n=3$ links,

$$\boldsymbol{H} = \sum_{i=1}^{3} \boldsymbol{H}_i. \tag{6.14}$$

The inertia matrices $\boldsymbol{H}_i$ for the individual links are derived below. The notation $c_{ij} = \cos(q_i+q_j)$ is used. The following trigonometric relationships are used to simplify intermediate matrices:

$$\begin{aligned} \sin(\alpha)\cdot\sin(\beta) &= 0.5\cdot(\cos(\alpha-\beta)-\cos(\alpha+\beta)) \\ \cos(\alpha)\cdot\cos(\beta) &= 0.5\cdot(\cos(\alpha-\beta)+\cos(\alpha+\beta)). \end{aligned} \tag{6.15}$$

Arm link 1:

$$H_1 = \begin{bmatrix} m_1\cdot l_{m2} & 0 & 0 \\ 0 & 0 & 0 \\ 0 & 0 & 0 \end{bmatrix} \tag{6.16}$$

Arm link 2:

$$H_2 = \begin{bmatrix} m_2(l_1^2+l_{m2}^2+2l_1l_{m2}c_2) & m_2(l_{m2}^2+l_1l_{m2}c_2) & 0 \\ m_2(l_{m2}^2+l_1l_{m2}c_2) & m_2l_{m2}^2 & 0 \\ 0 & 0 & 0 \end{bmatrix} \tag{6.17}$$

Arm link 3:

$$H_3 = \begin{bmatrix} h_{3,11} & h_{3,12} & h_{3,13} \\ h_{3,21} & h_{3,22} & h_{3,23} \\ h_{3,31} & h_{3,32} & h_{3,33} \end{bmatrix}, \tag{6.18}$$

with

$$\begin{aligned} h_{3,11} &= m_3(l_1^2+l_2^2+l_{m3}^2+2l_1l_2c_2+2l_1l_{m3}c_{23}+2l_2l_{m3}c_3) \\ h_{3,12} &= m_3(l_2^2+l_{m3}^2+l_1l_2c_2+l_1l_{m3}c_{23}+2l_2l_{m3}c_3) \\ h_{3,13} &= m_3(l_1l_{m3}c_{23}+l_2l_{m3}c_3) \\ h_{3,21} &= h_{3,12} \\ h_{3,22} &= m_3(l_2^2+l_{m3}^2+2l_2l_{m3}c_3) \\ h_{3,23} &= m_3(l_{m3}^2+l_2l_{m3}c_3) \\ h_{3,31} &= h_{3,13} \\ h_{3,32} &= h_{3,23} \\ h_{3,33} &= m_3l_{m3}^2. \end{aligned} \tag{6.19}$$

Finally, the elements of $\boldsymbol{H}$ in Equation (6.14) are $h_{ij} = h_{1,ij} + h_{2,ij} + h_{3,ij}$:

$$\begin{aligned}
h_{11} &= m_1 l_{m2} + m_2(l_1^2 + l_{m2}^2 + 2l_1 l_{m2} c_2) + \\
&\quad m_3(l_1^2 + l_2^2 + l_{m3}^2 + 2l_1 l_2 c_2 + 2l_1 l_{m3} c_{23} + 2l_2 l_{m3} c_3) \\
h_{12} &= m_2(l_{m2}^2 + l_1 l_{m2} c_2) + m_3(l_2^2 + l_{m3}^2 + l_1 l_2 c_2 + l_1 l_{m3} c_{23} + 2l_2 l_{m3} c_3) \\
h_{13} &= m_3(l_1 l_{m3} c_{23} + l_2 l_{m3} c_3) \\
h_{21} &= m_2(l_{m2}^2 + l_1 l_{m2} c_2) + m_3(l_2^2 + l_{m3}^2 + l_1 l_2 c_2 + l_1 l_{m3} c_{23} + 2l_2 l_{m3} c_3) \\
h_{22} &= m_2 l_{m2}^2 + m_3(l_2^2 + l_{m3}^2 + 2l_2 l_{m3} c_3) \\
h_{23} &= m_3(l_{m3}^2 + l_2 l_{m3} c_3) \\
h_{31} &= h_{13} \\
h_{32} &= h_{23} \\
h_{33} &= m_3 l_{m3}^2.
\end{aligned} \tag{6.20}$$

The $\boldsymbol{C}$ matrix of centripetal and Coriolis torques is obtained using Equation (6.11):

$$\begin{aligned}
c_{11} &= \sum_{k=1}^{n} (c_{11,k})\dot{q}_k = \sum_{k=1}^{n} \frac{1}{2}\left(\frac{\partial h_{11}}{\partial q_k} + \frac{\partial h_{1k}}{\partial q_1} - \frac{\partial h_{1k}}{\partial q_1}\right)\dot{q}_k = \frac{1}{2}\sum_{k=1}^{n} \frac{\partial h_{11}}{\partial q_k}\dot{q}_k \\
&= -(m_2 l_1 l_{m2} s_2 + m_3(l_1 l_2 s_2 + l_1 l_{m3} s_{23}))\dot{q}_2 - m_3(l_1 l_{m3} s_{23} + l_2 l_{m3} s_3)\dot{q}_3,
\end{aligned} \tag{6.21}$$

with

$$\begin{aligned}
k = 1&: \frac{\partial h_{11}}{\partial q_1} = 0 \\
k = 2&: \frac{\partial h_{11}}{\partial q_2} = -2m_2 l_1 l_{m2} s_2 - 2m_3(l_1 l_2 s_2 + l_1 l_{m3} s_{23}) \\
k = 3&: \frac{\partial h_{11}}{\partial q_3} = -2m_3(l_1 l_{m3} s_{23} + l_2 l_{m3} s_3);
\end{aligned}$$

$$\begin{aligned}
c_{12} &= \sum_{k=1}^{n} (c_{12,k})\dot{q}_k = \sum_{k=1}^{n} \frac{1}{2}\left(\frac{\partial h_{12}}{\partial q_k} + \frac{\partial h_{1k}}{\partial q_2} - \frac{\partial h_{2k}}{\partial q_1}\right)\dot{q}_k \\
&= -(m_2 l_1 l_{m2} s_2 + m_3(l_1 l_2 s_2 + l_1 l_{m3} s_{23}))\dot{q}_1 \\
&\quad -(m_2 l_1 l_{m2} s_2 + m_3(l_1 l_2 s_2 + l_1 l_{m3} s_{23}))\dot{q}_2 - m_3(l_1 l_{m3} s_{23} + l_2 l_{m3} s_3)\dot{q}_3,
\end{aligned} \tag{6.22}$$

with

$$
\begin{aligned}
k = 1: & \frac{\partial h_{12}}{\partial q_1} = \frac{\partial h_{21}}{\partial q_1} = 0 \\
& \frac{\partial h_{11}}{\partial q_2} = \text{(see above)} \\
k = 2: & \ 2\frac{\partial h_{12}}{\partial q_2} = 2(-m_2 l_1 l_{m2} s_2 - m_3(l_1 l_2 s_2 + l_1 l_{m3} s_{23})) \\
& \frac{\partial h_{22}}{\partial q_1} = 0 \\
k = 3: & \frac{\partial h_{12}}{\partial q_3} = -m_3(l_1 l_{m3} s_{23} + 2 l_2 l_{m3} s_3) \\
& \frac{\partial h_{13}}{\partial q_2} = -m_3 l_1 l_{m3} s_{23} \\
& \frac{\partial h_{23}}{\partial q_1} = 0;
\end{aligned}
$$

$$
\begin{aligned}
c_{13} = & \sum_{k=1}^{n} (c_{13,k})\dot{q}_k = \sum_{k=1}^{n} \frac{1}{2}\left(\frac{\partial h_{13}}{\partial q_k} + \frac{\partial h_{1k}}{\partial q_3} - \frac{\partial h_{3k}}{\partial q_1}\right)\dot{q}_k \\
= & -(m_3(l_1 l_{m3} s_{23} + l_2 l_{m3} s_3))\dot{q}_1 - (m_3(l_1 l_{m3} s_{23} + l_2 l_{m3} s_3))\dot{q}_2 \\
& -(m_3(l_1 l_{m3} s_{23} + l_2 l_{m3} s_3))\dot{q}_3,
\end{aligned}
\tag{6.23}
$$

with

$$
\begin{aligned}
k = 1: & \frac{\partial h_{13}}{\partial q_1} = 0 \\
& \frac{\partial h_{11}}{\partial q_3} = \text{(see above)} \\
& \frac{\partial h_{31}}{\partial q_1} = \frac{\partial h_{13}}{\partial q_1} = 0 \\
k = 2: & \frac{\partial h_{13}}{\partial q_2} = -m_3 l_1 l_{m3} s_{23} \\
& \frac{\partial h_{12}}{\partial q_3} = \text{(see above)} \\
& \frac{\partial h_{32}}{\partial q_1} = \frac{\partial h_{23}}{\partial q_1} = 0 \\
k = 3: & \ 2\frac{\partial h_{13}}{\partial q_3} = -2m_3(l_1 l_{m3} s_{23} + l_2 l_{m3} s_3) \\
& \frac{\partial h_{33}}{\partial q_1} = 0;
\end{aligned}
$$

$$
c_{21} = c_{12}; \tag{6.24}
$$

$$c_{22} = \sum_{k=1}^{n} (c_{22,k})\dot{q}_k = \sum_{k=1}^{n} \frac{1}{2}\left(\frac{\partial h_{22}}{\partial q_k} + \frac{\partial h_{2k}}{\partial q_2} - \frac{\partial h_{2k}}{\partial q_2}\right)\dot{q}_k = \frac{1}{2}\sum_{k=1}^{n} \frac{\partial h_{22}}{\partial q_k}\dot{q}_k, \quad (6.25)$$

$$= -(m_3 l_2 l_{m3} s_3)\dot{q}_3$$

with

$$k = 1{:}\, \frac{\partial h_{22}}{\partial q_1} = 0$$

$$k = 2{:}\, \frac{\partial h_{22}}{\partial q_2} = 0$$

$$k = 3{:}\, \frac{\partial h_{22}}{\partial q_3} = -2m_3 l_2 l_{m3} s_3;$$

$$c_{31} = c_{13}; \quad (6.26)$$

$$c_{33} = \sum_{k=1}^{n} (c_{33,k})\dot{q}_k = \sum_{k=1}^{n} \frac{1}{2}\left(\frac{\partial h_{33}}{\partial q_k} + \frac{\partial h_{3k}}{\partial q_3} - \frac{\partial h_{3k}}{\partial q_3}\right)\dot{q}_k = \frac{1}{2}\sum_{k=1}^{n} \frac{\partial h_{33}}{\partial q_k}\dot{q}_k = 0, \quad (6.27)$$

with

$$k = 1{:}\, \frac{\partial h_{33}}{\partial q_1} = 0$$

$$k = 2{:}\, \frac{\partial h_{33}}{\partial q_2} = 0$$

$$k = 3{:}\, \frac{\partial h_{33}}{\partial q_3} = 0.$$

The gravity force vector $\boldsymbol{G}$ is computed using Equation (6.12). The constant gravity vector in the base frame is $\boldsymbol{g} = \begin{bmatrix} 0 & 0 & -9.81\text{m}\cdot\text{s}^{-2} \end{bmatrix}^T$. With the general 6×3 Jacobian $\boldsymbol{J}_L(\theta)$ in Equation (5.48) the vector elements of $\boldsymbol{G}$ are zero:

$$G_i = \sum_{i=1}^{3} m_i \boldsymbol{g}^T \boldsymbol{j}_{L,i} = 0 \quad (6.28)$$

because the third row of $\boldsymbol{J}_L(\theta)$ is zero. Gravitational forces have no effect on the arm, which moves in a horizontal plane, perpendicular to the gravity field. Therefore $\boldsymbol{G} = \boldsymbol{0}$.

The vector of generalized forces in Equation (6.13) comprises only motor torques acting on the revolute joints, thus $\boldsymbol{Q} = \tau$. The end-effector of the substrate-handling robot does not make contact with the environment, and external forces are not present. The joint space dynamical model for the SCARA-type robot arm in Figure 6.1 is therefore

$$\boldsymbol{H}(\theta)\ddot{\theta} + \boldsymbol{C}(\theta, \dot{\theta})\dot{\theta} = \tau. \quad (6.29)$$

The model does not include friction terms. □

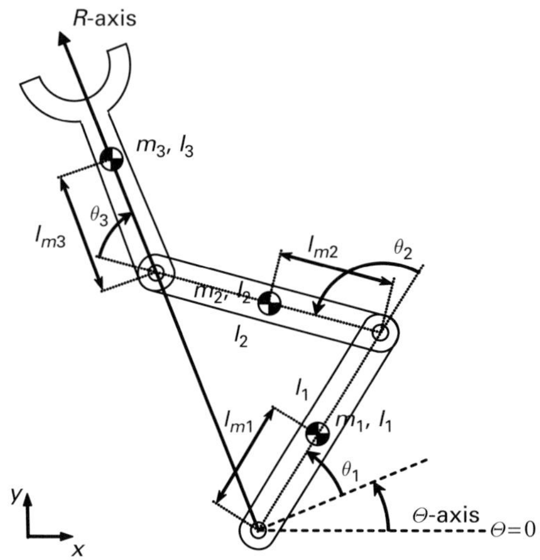

Figure 6.1 Dynamic parameters of a SCARA-type robot arm (Example 6.1).

6.1.3 System identification

System identification is a discipline that uses mathematical tools and algorithms to build mathematical models of dynamic systems based on measured data from the physical system. See (Ljung, 1987) for a comprehensive discussion. Several software tools are available that can be used to improve an existing dynamical robot model. For example, viscous and static friction models can be added to the general dynamic model in Equation (6.1):

$$\boldsymbol{H}(\boldsymbol{q})\ddot{\boldsymbol{q}} + \boldsymbol{C}(\boldsymbol{q}, \dot{\boldsymbol{q}})\dot{\boldsymbol{q}} + \boldsymbol{F}_v\dot{\boldsymbol{q}} + \boldsymbol{F}_s\mathrm{sgn}(\dot{\boldsymbol{q}}) + \boldsymbol{G}(\boldsymbol{q}) = \boldsymbol{Q}. \tag{6.30}$$

$\boldsymbol{F}_v \in \mathbb{R}^{n\times n}$ and $\boldsymbol{F}_s \in \mathbb{R}^{n\times n}$ are diagonal matrices of viscous and static friction coefficients, respectively, and the vector $\mathrm{sgn}(\dot{\boldsymbol{q}}) \in \mathbb{R}^{n\times 1}$ comprises the sign functions of joint velocities. The modeled friction forces and torques in Equation (6.30) improve the model accuracy if moving parts with a relatively high friction are used in robot drive trains. $\boldsymbol{F}_v$ and $\boldsymbol{F}_s$ can be determined with system identification algorithms and measured data. Refer to the literature for further details, for example (Ljung, 1987; Sciavicco and Siciliano, 1996).

6.2 Robot motion control

Many robot control methods and controller architectures are available. This section provides a brief overview of control methods for robots. Refer to the robotics literature for further details, for example (Asada and Slotine, 1986; Åström and Hägglund, 1994; Craig, 1989; Hristu-Varsakelis and Levine, 2005; Mohler, 1991; Sciavicco and Siciliano, 1996; Spong and Vidyasagar, 1989; Younkin, 1996).

6.2.1 Two feedback control concepts

Figure 6.2 shows the two control system concepts used for robot control: (a) operational space control, and (b) joint space control. The block arrows indicate vector signals. The following notation is used:

$\boldsymbol{x}_{cmd} \in \mathbb{R}^{m\times 1}$: commanded end-effector pose in workspace
$\boldsymbol{x} \in \mathbb{R}^{m\times 1}$: actual, measured end-effector pose in workspace
$\boldsymbol{e} \in \mathbb{R}^{m\times 1}$: end-effector position error
$\boldsymbol{u} \in \mathbb{R}^{n\times 1}$: control signal computed by the robot controller
$\theta_{cmd} \in \mathbb{R}^{n\times 1}$: commanded joint displacements in joint space
$\theta \in \mathbb{R}^{n\times 1}$: actual, measured joint displacements in joint space
$\tau \in \mathbb{R}^{n\times 1}$: actuator signal, here the commanded motor torques.

Figure 6.2a shows a basic robot control system architecture with negative feedback and a robot as the controlled plant. Given an end-effector pose (position and orientation) trajectory in the robot's workspace, the control objective is to minimize the error $\boldsymbol{e} = \boldsymbol{x}_{cmd} - \boldsymbol{x}$. The control signal $\boldsymbol{u}$ is computed to achieve this goal. Figure 6.2a illustrates one of the disadvantages of the operational space control concept: it requires measurement of the actual end-effector position and orientation $\boldsymbol{x}$, which is usually a non-trivial, sometimes impractical task.

Figure 6.2b outlines the controller architecture for the joint space control scheme. It is not the end-effector position in workspace which is measured, but the joint displacements in joint space.

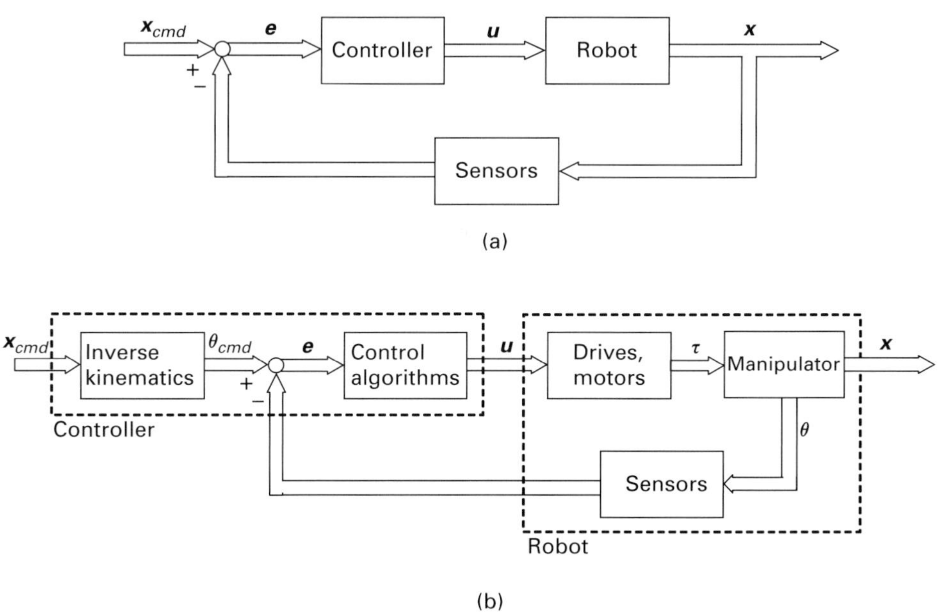

Figure 6.2 Feedback control architecture in (a) operational space, (b) joint space.

The following assumptions apply to the majority of substrate-handling applications and facilitate the design of suitable control systems:

- The end-effector pose for loading and unloading substrates is determined by the constant reference location of the substrate transfer chuck, for example.
- The required manipulator positioning repeatability can be achieved with common mechanical designs and manufacturing techniques.
- Point-to-point control along straight paths, to and from fixed reference points in workspace, is sufficient: only the initial and final positions of a motion path are critical.
- Robot motion in free space is needed. There is no interaction between the end-effector and the environment.
- The dynamic interaction between joints may be disregarded and individual joints can be controlled independently.
- Gravitational forces do not apply because the SCARA-type arms move in horizontal planes.

The fixed reference points, the loading and unloading positions, are kinematically calibrated ('taught'). A sufficient and known robot positioning repeatability (Section 7.3) allows the use of the joint space control concept, that is, controlling the robot without measuring the end-effector position. Only the joint displacements $\boldsymbol{\theta}$ are known. The resulting manipulator pose $\boldsymbol{x}$ in workspace is inferred from the forward kinematics model. This control method tends to be more practical than operational space control: in substrate handling a common requirement is sufficient positioning repeatability at specified, taught reference points. Therefore joint space control may be used, because positioning feedback at the end-effector is not required. Mechanical solutions, as described in Example 5.6 for straight paths, are also suitable.

The controller design is further facilitated by the assumptions that there is no contact between the end-effector and the robot's environment, and that the dynamic interaction between joints may be disregarded for high gear ratios. Light-weight arm links also reduce the effect of dynamic interaction. This permits the independent control of individual joints and leads to a decentralized control system architecture, discussed in Section 6.2.2.

6.2.2 Decentralized control of individual joints

The decentralized control of individual joints facilitates the controller design and software development. Given the above simplifications a manipulator with n independent joints utilizes n controllers, as such forming n single-input/single-output (SISO) systems. A supervisor controller coordinates the tracking of commanded trajectories and other high-level tasks. A generic controller for an individual revolute joint is described below, including a dynamical model of the rotary electric DC motor and the link load. Then a specific controller type is selected as an example. Classical frequency domain methods are used.

Dynamical model of an individual joint and arm link

The following notation is used:

v_a: applied armature voltage (V)
i_a: armature current (A)

R_a: armature resistance (Ω)
L_a: armature inductance (H)
v_{emf}: motor back EMF (V)
τ_m: motor torque (N·m)
θ_m: angular displacement of the motor shaft (rad)
J_m: motor inertia (kg·m^2)
J_l: load inertia (kg·m^2)
N: gear ratio from motor to load (unitless)
b: damping ratio (friction coefficient) of the motor shaft (N·m·(rad/s)$^{-1}$ = N·m·rad^{-1}·s = N·m·s)
τ_d: external torque disturbance applied at the motor shaft (N·m)
G_v: voltage gain of the power amplifier (unitless)
K_τ: motor torque constant (N·m·A^{-1})
K_e: motor back EMF constant (V·(rad/s)$^{-1}$ = V·rad^{-1}·s = V·s)
K_m: motor gain constant ((rad/s)·V^{-1} = rad·s^{-1}·V^{-1} = V^{-1}·s^{-1}))
T_m: motor time constant (s).

Figure 6.3a shows a schematic diagram of an armature-controlled rotary DC motor and the arm link load. The drive train from motor to the link is geared, which reduces the reflected inertia (the load inertia applied to the motor) to J_l/N^2. The external torque disturbance τ_d represents any unmodeled, nonlinear dynamic interaction between the manipulator joints.

The motor constants K_τ and K_e have the same values if specified in SI units. This can be tested using their measurement units: $[K_e] = \mathrm{V}\cdot\mathrm{s} = \mathrm{kg}\cdot\mathrm{m}^2\cdot\mathrm{s}^{-3}\cdot\mathrm{A}^{-1}\cdot\mathrm{s} = \mathrm{N}\cdot\mathrm{m}\cdot\mathrm{A}^{-1} = [K_\tau]$.

Figure 6.3b shows a block diagram of transfer functions for the schematic diagram in Figure 6.3a, including the first-order system models of the armature circuit, the manipulator inertia, and viscous friction. The disturbance τ_d is reduced to τ_d/N by the transmission before reaching the motor. A gear ratio $N \gg 1$ tends to linearize the motor and arm link dynamics, and therefore the SISO control problem.

The motor torque is given by

$$\tau_m = K_\tau i_a. \tag{6.31}$$

The back EMF is a function of the motor speed:

$$v_{emf} = K_e \frac{d\theta_m}{dt}. \tag{6.32}$$

Two differential equations are used to describe rotary electric DC motors. The torque balance equation for the motor shaft is

$$J\frac{d^2\theta_m}{dt^2} + b\frac{d\theta_m}{dt} - \frac{1}{N}\tau_d = K_\tau i_a, \text{ with } J = J_m + \frac{J_l}{N^2}. \tag{6.33}$$

J is the total inertia at the motor shaft. The second differential equation is based on Kirchhoff's voltage law (see Figure 6.3a),

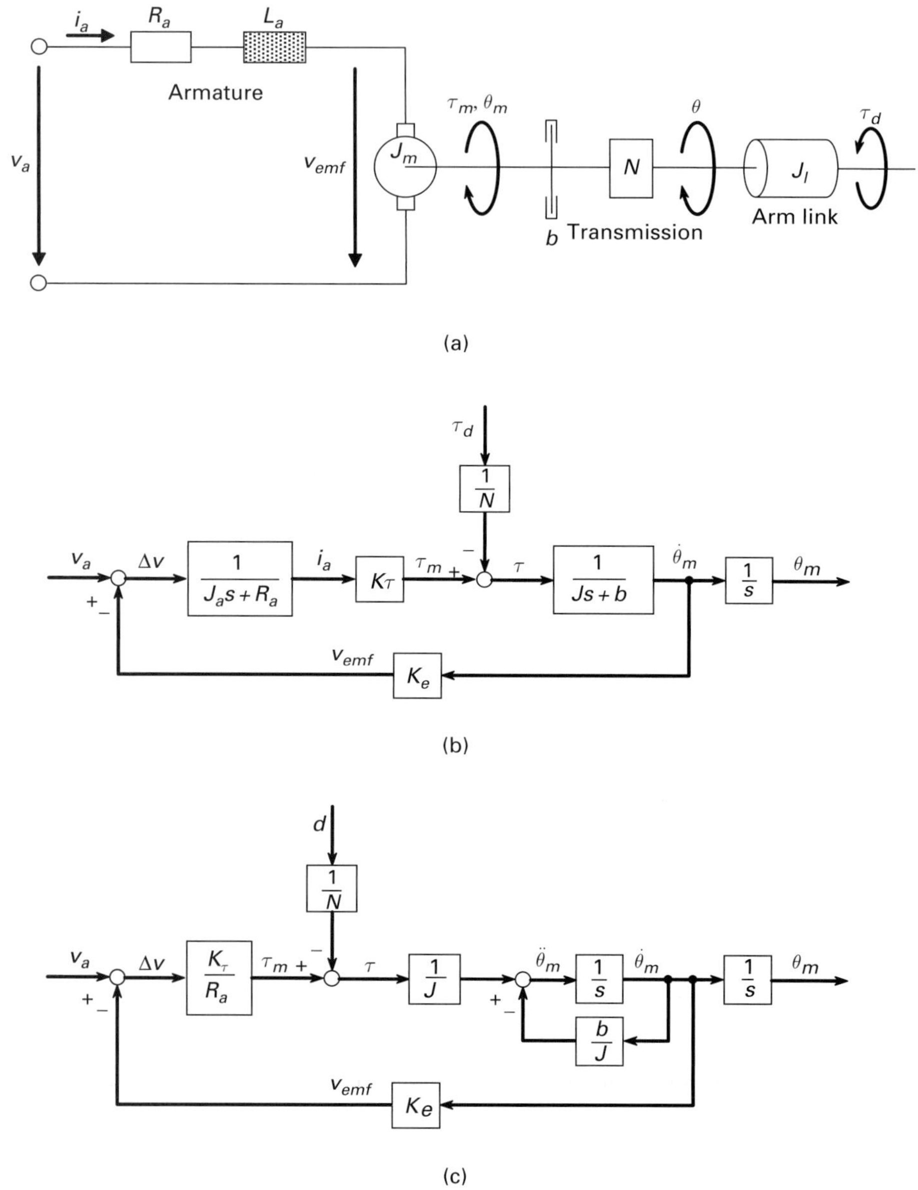

Figure 6.3 (a) Schematic diagram of a DC motor with load and transmission; (b) block diagram with modeled disturbance; (c) simplified block diagram ($L_a = 0$).

$$L_a \frac{di_a}{dt} + R_a i_a = v_a - K_e \frac{d\theta_m}{dt}. \tag{6.34}$$

The transfer function for the motor and load can then be obtained with the Laplace transforms of Equations (6.33) and (6.34):

$$J\Theta_m(s)s^2 + b\Theta_m(s)s = K_\tau I_a(s) \tag{6.35}$$

$$L_a I_a(s)s + R_a I_a(s) = V_a(s) - K_e \Theta_m(s)s. \tag{6.36}$$

The armature current is

$$I_a(s) = \frac{\Theta_m(s)s}{K_\tau}(Js + b). \tag{6.37}$$

The dependence on the complex frequency s is omitted below for simplicity. The electric current is eliminated from Equation (6.36) using Equation (6.37):

$$\frac{\Theta_m s}{K_\tau}(J\,s + b)\,(L_a\,s + R_a) = V_a - K_e\,\Theta_m\,s. \tag{6.38}$$

The transfer function $P(s)$ of the individual arm joint (the plant to be controlled) from armature voltage to the joint displacement, is

$$P(s) = \frac{\Theta_m(s)}{V_a(s)} = \frac{K_\tau}{s((Js + b)(L_a s + R_a) + K_e K_\tau)}. \tag{6.39}$$

L_a is usually small and may be disregarded:

$$\begin{aligned} P(s) &= \frac{1}{s}\frac{K_\tau}{R_a J s + (R_a b + K_e K_\tau)} \\ &= \frac{1}{s}\frac{K_m}{T_m s + 1}, \end{aligned} \tag{6.40}$$

with

$$K_m = \frac{K_\tau}{bR_a + K_e K_\tau} \tag{6.41}$$

$$T_m = \frac{JR_a}{bR_a + K_e K_\tau}. \tag{6.42}$$

The block diagram for $P(s)$ in Equation (6.40) is shown in Figure 6.3c.

A convenient method for quickly verifing a derived system model like $P(s)$ is a measurement unit test. For example, the units of the total torque and the angular acceleration are (N·m) and ($1/s^2$), respectively. This is tested using Equations (6.31) and (6.33) below.

With $\dfrac{d\theta_m}{dt} = \dot{\theta}_m, \dfrac{d^2\theta_m}{dt^2} = \ddot{\theta}_m$, and $\dfrac{di_a}{dt} = \dot{i}_a$,

$$[\tau_m] = [K_\tau i_a] = [K_\tau \frac{V_a - V_{emf}}{R_a}] = \frac{\mathrm{N\cdot m}}{\mathrm{A}}\frac{\mathrm{V}}{\Omega} = \mathrm{N\cdot m} \tag{6.43}$$

$$[\ddot{\theta}_m] = [\left(\frac{\tau_m}{J} + \frac{\tau_d}{NJ}\right) - \frac{b}{J}\dot{\theta}_m] = \frac{\mathrm{N\cdot m}}{\mathrm{kg\cdot m^2}} - \frac{\mathrm{N\cdot m}}{\mathrm{s^{-1}\cdot kg\cdot m^2}}\frac{1}{\mathrm{s}} = \frac{\mathrm{rad}}{\mathrm{s^2}}. \tag{6.44}$$

The resulting units are correct. Similarly, the units of the motor constant (speed per voltage) and the motor time constant (time) in Equations (6.41) and (6.42) are also correct:

$$[K_m] = \frac{\mathrm{N \cdot m \cdot A^{-1}}}{\mathrm{N \cdot m \cdot s} \cdot \Omega + \mathrm{V \cdot s \cdot N \cdot m \cdot A^{-1}}} = \mathrm{V^{-1} \cdot s^{-1}} \tag{6.45}$$

$$[T_m] = \frac{\mathrm{kg \cdot m^2} \cdot \Omega}{\mathrm{N \cdot m \cdot s} \cdot \Omega + \mathrm{V \cdot s \cdot N \cdot m \cdot A^{-1}}} = \mathrm{s}, \text{ with } \Omega = \frac{V}{A}. \tag{6.46}$$

Therefore $P(s)$ in Equation (6.40) produces the correct physical quantity:

$$[P(s)] = [\frac{1}{s}\frac{K_m}{T_m s + 1}] = \frac{1}{\mathrm{s^{-1}}}\frac{\mathrm{V^{-1} \cdot s^{-1}}}{(\mathrm{s \cdot s^{-1}} + 1) =} \mathrm{rad \cdot V^{-1}} = \mathrm{V^{-1}} \tag{6.47}$$

'Motor angle per volt' is the expected unit for the transfer function of a voltage-controlled revolute joint. A measurement unit test is not a proof, but a convenient and quick test of mathematical models. A correct unit indicates that the underlying equation is probably correct.

Decentralized control of an independent joint

Many control strategies are available for robot manipulators. For the above revolute joint a decentralized, SISO control strategy is suitable, as was shown above for the independent joint model. The control strategy leads towards a bus-based, distributed implementation that offers flexibility and scalability. In the following a common and generic controller architecture is outlined. Then a specific controller, based on the generic architecture, is selected as an example.

Figure 6.4 shows a generic control systems architecture with nested, negative feedback loops for position (outermost loop), velocity, and acceleration (innermost loop). The plant $P(s)$ is an independent arm joint with an armature-controlled DC motor and the load as described by Equation (6.40). The position controller C_P, velocity controller C_V, and acceleration controller C_A receive feedback signals from the respective sensors,

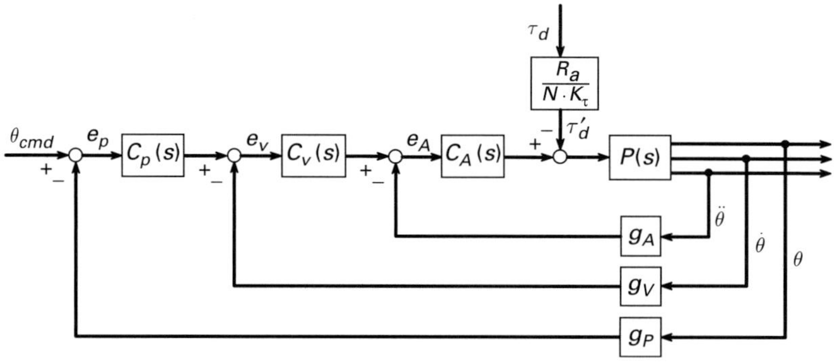

Figure 6.4 Generic controller for an individual arm joint with load and disturbance.

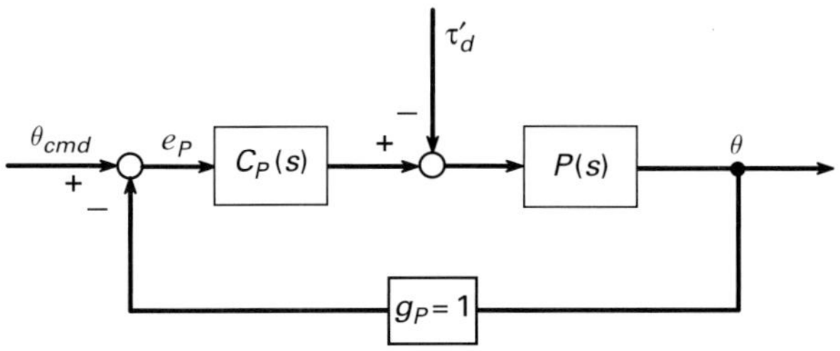

Figure 6.5 Position-controlled arm joint.

represented by the sensor constants g_P, g_V, and g_A. The amplifier gain is part of $P(s)$. The external torque disturbance τ_d has been moved from the motor shaft in Figure 6.3c to the plant input, which is compensated for by the inverse factor R_a/K_τ. Note that $\tau_d' = \tau_d \cdot R_a/K_\tau$ is a voltage.

The control objectives are: (a) to provide stability of the controlled arm joint; (b) to minimize the errors e_P, e_V, and e_A in the presence of τ_d. The control system architecture in Figure 6.4 offers several controller solutions. In the simplest case only position control is used, that is, $g_V = g_A = 0$ and $C_V = C_A = 1$. This particular solution is presented as an example below. The task is to design C_P that meets the above control objectives.

The position-controlled arm joint is reduced to the architecture in Figure 6.5, with $g_P = 1$.

The transfer function from θ_{cmd} to θ, from the desired to the actual joint displacement, is

$$G(s) = \frac{P(s)C_P(s)}{1 + P(s)C_P(s)}. \tag{6.48}$$

A suitable and frequently used position controller for a revolute arm joint is a proportional-integral (PI) controller:

$$\begin{aligned} C_P(s) &= K_P + \frac{K_I}{s} \\ &= K_C \frac{T_C s + 1}{s}, \end{aligned} \tag{6.49}$$

with $T_C = \dfrac{K_P}{K_I}$, $K_C = \dfrac{K_P}{T_C}$.

With the joint model P in Equation (6.40) and the position controller C_P in Equation (6.49) the open-loop transfer function is

$$P(s)C_P(s) = \frac{K_m K_C (T_C s + 1)}{s^2 (T_m s + 1)}. \tag{6.50}$$

The transfer function of the controlled joint is therefore

$$G(s) = \frac{K_C K_m (T_C s + 1)}{s^2 (T_m s + 1) + K_C K_m (T_C s + 1)} \tag{6.51}$$

$$= \frac{K_C K_m T_C s + K_C K_m}{s^3 + \frac{1}{T_m} s^2 + \frac{K_C K_m T_C}{T_m} s + \frac{K_C K_m}{T_m}}. \tag{6.52}$$

Stability

It is well known that the controlled joint in Equation (6.51) is stable if all poles of the transfer function are in the left-half plane of the complex s-plane. The root locus is a convenient tool for the stability analysis of linear, time-invariant (LTI) SISO systems. It determines and plots the closed-loop poles, the roots of the characteristic equation, as a function of the system's open-loop gain. A suitable controller gain that places all poles in the left-half plane and achieves the desired dynamics can then be selected from the root locus (Kuo, 1981; Ogata, 1990).

The characteristic equation for the closed loop is the denominator of the closed-loop transfer function in Equation (6.48), $1 + P(s)C_P(s)$. It is represented with the adjustable gain K_C as a multiplying factor and then set to zero to determine its roots:

$$1 + K_C P(s) C'_P(s) = 0, \quad \text{with} \quad C'_P(s) = \frac{T_C s + 1}{s}, \tag{6.53}$$

or, with Equation (6.40),

$$\begin{aligned} K_C P(s) C'_P(s) &= K_C \frac{K_m}{s(T_m s + 1)} \frac{T_C s + 1}{s} \\ &= -1. \end{aligned} \tag{6.54}$$

With the changing gain $-\infty < K_C < \infty$ the graph of the root loci begins and ends at the poles and zeros of the open-loop system $K_C P(s) C'_P(s)$, respectively. The controlled plant in Equation (6.50) has three poles, a double pole at the origin and one pole at $-1/T_m$. Its zero is at $-1/T_C$. All other root loci are determined using the characteristic equation in Equation (6.54), which is divided into the phase condition and magnitude condition:

$$\begin{aligned} \angle\big(P(s)C'_P(s)\big) &= \sum_{i=1}^{m} \angle(s + z_i) - \sum_{j=1}^{n} \angle(s + p_i) \\ &= (2k+1) \cdot 180^\circ, \ 0 \le K_C < \infty, \ k = 0, \pm 1, \pm 2, \ldots \end{aligned} \tag{6.55}$$

$$\big|K_C P(s) C'_P(s)\big| = K_C \frac{\prod_{i=1}^{m} |s + z_i|}{\prod_{j=1}^{n} |s + p_i|} = 1, \ 0 < K_C < \infty. \tag{6.56}$$

A root locus ('locus' is Latin for 'location') is determined using the phase condition in Equation (6.55). The sum of all phases of poles p_i and zeros z_i must be a multiple of 180°. In order to obtain a particular root locus, either a single real or a complex conjugated pair of poles, the controller gain K_C may be determined using the amplitude condition in Equation (6.56). Several software tools are available for these computations and for

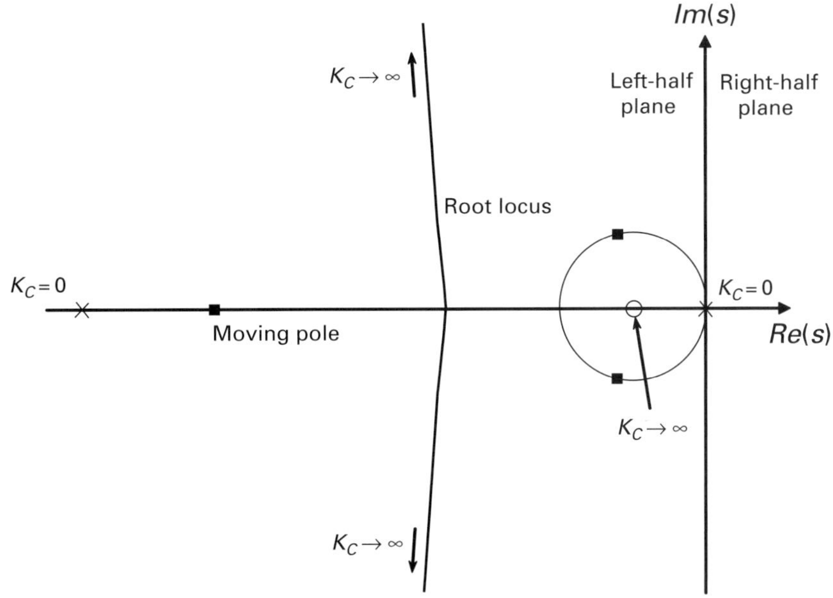

Figure 6.6 Root locus plot for position-controlled arm joint in Example 6.2.

plotting the root locus curves. The analysis in the example below shows that a stable position control for an individual revolute joint requires $T_C \gg T_m$ (Figure 6.6).

Disturbance rejection (steady-state)

The transfer function from the torque disturbance τ_d' to the actual joint displacement θ of the control system in Figure 6.5 is

$$G_\tau(s) = -\frac{P(s)}{1 + P(s)C_P(s)} \tag{6.57}$$

$$= -\frac{K_m s}{s^3 + \dfrac{1}{T_m}s^2 + \dfrac{K_C K_m T_C}{T_m}s + \dfrac{K_C K_m}{T_m}}. \tag{6.58}$$

$G_\tau(s)$, and $G(s)$ in Equation (6.52), have the same characteristic equations. Equation (6.58) suggests that increasing the controller gain K_C reduces the disturbance effects during the system's transient. However, a high gain can cause oscillations so a balance must be found. Equation (6.58) is used to determine the steady-state disturbance rejection of the PI-controlled system. It can be shown, with the final value property of the Laplace transform, that the steady state error e_{ss} due to τ_d' is zero:

$$\begin{aligned}
e_{ss} &= \lim_{t\to\infty} e(t) \\
&= \lim_{t\to\infty} (u(t) * g_{\tau_d}(t)) \\
&= \lim_{s\to 0} s\left(\frac{1}{s} G_{\tau_d}(s)\right) \\
&= \lim_{s\to 0} s\frac{1}{s}\left(-\frac{K_m s}{s^3 + \frac{1}{T_m}s^2 + \frac{K_C K_m T_C}{T_m}s + \frac{K_C K_m}{T_m}}\right) = 0.
\end{aligned} \tag{6.59}$$

Here $u(t)$ is the step function as a function of time t, and * denotes the convolution. In summary, the PI controller is a suitable choice for stability, steady-state disturbance rejection, and for designing a decentralized control architecture for individual revolute joints. However, additional performance criteria may apply in controller design specifications, most commonly rise and settling times, and overshoot. Depending on the overall set of criteria, controller types other than the PI type may be more suitable.

Example 6.2: PI control for one revolute joint

The qualitative root locus curve for an individual joint is shown in Figure 6.6. Consider the following motor parameters used in Equation (6.40): $J = 0.01\ \mathrm{kg}\cdot\mathrm{m}^2\mathrm{s}^{-2}$, $b = 0.1\ \mathrm{N}\cdot\mathrm{m}\cdot\mathrm{s}$, $K = K_t = K_e = 0.01\ \mathrm{N}\cdot\mathrm{m}\cdot\mathrm{A}^{-1}$, $R_a = 1\,\Omega$, $L_a = 0.5\ \mathrm{H}$. The PI controller in Equation (6.49) has the time constant $T_C = 10T_m$.

The open-loop (K_C=0) poles and zeros are represented by × and o, respectively. The root loci, one real pole and one pair of complex conjugated poles, move as a function of the adjustable controller gain K_C and are denoted by solid squares (■). □

6.3 Networked, decentralized robot control

The above discussion about the control of individual manipulator joints has shown that networked control systems can be suitable for the coordinated and decentralized motion control of multi-axis robots. This section discusses the basic requirements for such control systems.

6.3.1 Traditional controller architecture

Traditional, centralized controller architectures for substrate-handling robots combine all required subsystems in one central device:

- A microprocessor for supervising and coordinating various controller functions
- A motion-control board for computing the control signals for all motors based on user commands and sensor feedback
- Power amplifiers that convert the low-voltage control signal to the appropriate motor current

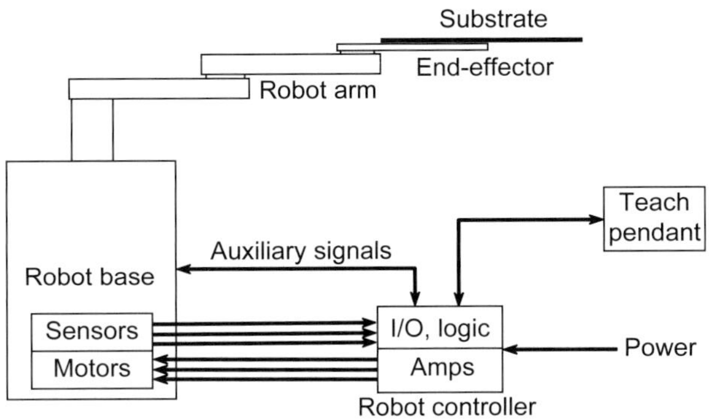

Figure 6.7 Traditional architecture with external robot controller and a three-axis robot.

- An input/output (I/O) interface for peripherals, for example 'teach' pendants and an emergency-off switch
- A power supply
- A data bus that connects the above subsystems.

Figure 6.7 shows a typical system configuration: the central controller is external to the robot. Auxiliary signals can be used for sensor signals from the end-effector, for example for 'substrate present' and 'gripper closed' signals; or to turn on vacuum suction at the end-effector for holding substrates. Teach pendants, and more recently laptop computers, are the main human/robot interface. Early commercial controllers used an ISA or EISA bus and a microprocessor with an 80x86 CPU, that is, the original computing platform of personal computers.

A disadvantage of the centralized controller architecture is that multiple, expensive cleanroom cables are needed to power motors and feed back sensor signals. In addition to the cost, this can contribute to particle contamination and create antenna effects and related EMC problems. Furthermore, not all motion-control cards and their macro languages support structured programming, and these offer only limited mathematical instruction. A stand-alone, external robot controller increases the footprint of a robot system, which is not desirable given the limited space in modern manufacturing tools. Section 6.3.2 discusses how bus-based control architectures can solve some or all of these limitations.

6.3.2 Realization of networked, decentralized control

Networked robot control is made possible by the availability of open bus standards and low-cost electronic components that provide deterministic data transfer at the required bandwidths. Bus-based, decentralized control of robots has advantages over the traditional centralized approach. First, the modularity and scalability of networked robot control facilitates the reconfiguration and re-use of control systems, and can therefore support different robot types with different numbers of axes. Such a unified controller platform in turn reduces the overall cost of a controller product. Second, a 'single-wire' (network) solution reduces costs by replacing several expensive cleanroom cables, which also increases reliability by reducing the risk of cable failures and manufacturing mistakes. Third, open architectures allow structured programming and the use of high-level programming languages. Many motion-control cards and their (often proprietary) macro languages have limitations in this respect. In addition to motion-control applications, networked control also enables remote diagnostics, troubleshooting, and controller tuning. Fourth, a decentralized controller with single-axis, smart motor amplifiers can be integrated with the robot, which reduces the footprint of the robot system. Following are examples of recent technology advances that support the above capabilities:

- Standard, open-architecture control networks provide sufficiently high update rates at low cost.
- Miniature servo amplifiers with high power densities of up to 0.04 W·mm^{-3} (continuous) and 0.06 W·mm^{-3} (peak) for one axis of motion are now available.
- 'Smart' amplifiers execute local, low-level control algorithms (usually PID-type algorithms) and report back to a supervisor controller.
- High-level programming languages for modern motion controllers offer structured programming and a high level of hardware abstraction.
- Advances in the theory of networked control systems provide a well-established theoretical foundation for networked, decentralized robot control (Hristu-Varsakelis and Levine, 2005).

Several commercial products are available that support a decentralized architecture. Bus standards for connecting electronic control units in a deterministic control network include the controller area network (CAN) bus with its CANopen protocol (Pfeiffer *et al.,* 2003); EtherCAT, an adaptation of the Ethernet for control automation technology; and Ethernet POWERLINK. Example 6.3 outlines the hardware and software architecture of network control for a three-axis robot using the CAN bus and CANopen standards. CANopen is an international-standard (EN 50325-4), higher-layer protocol for embedded control systems based on the CAN bus. Example 6.4 describes the application of CAN-bus architecture to the control of a dual-arm vacuum robot. Each arm is a closed kinematic chain with two driven revolute joints coordinated by a synchronized, networked control system, and achieves smooth, accurate motion trajectories.

Example 6.3: robot control using the CAN-bus standard

A CAN bus can transfer data at rates of up to 1 Mbit·s^{-1} at network lengths up to 40 m, which exceeds the requirements of the robot applications considered here. Figure 6.8a illustrates a control network for a three-axis robot using the CAN bus and CANopen standards. This architecture has also been extended to a five-axis robot with two closed kinematic chain arms (see Example 6.4). The supervisor controller and the single-axis servo amplifiers in Figure 6.8a can be placed inside the robot to reduce both its footprint and the total cable length. Each motor is equipped with a mechanical brake. The human operator coordinates the robot system using the host PC. For example, the operator assigns the robot's motion sequence within a manufacturing tool, which is then communicated to the supervisor controller via an Ethernet link. The supervisor controller then performs the high-level control tasks, for example trajectory planning and motion start/stop. It communicates with the single-axis, smart amplifiers using a CAN bus and the CANopen protocol. Each 'smart' amplifier includes a PID controller that executes the low-level, closed-loop control, thereby reducing the data traffic across the CAN bus.

A block diagram of the software architecture that operates on the hardware platform is shown in Figure 6.8b. The architecture is organized in several software layers with an increasing level of abstraction towards the graphical user interface (GUI). The GUI resides on the host PC. The following software runs on the supervisor controller:

- The robot application includes the kinematic and dynamic models and performs high-level tasks, for example generating the needed end-effector trajectories.
- The robot API (application programming interface) provides the interface to the robot application.
- The controller API provides the software interface to the low-level control functions of the smart amplifiers.
- The CANopen API provides the software interface to the CANopen protocol stack.
- The CANopen protocol stack provides the functions required for CAN-bus communication, network management, general system services, and device description.
- The CAN interface card API is the hardware interface to the CAN bus, and therefore provides access to the hardware layer of the network.
- The smart servo amplifiers are CAN-bus nodes equipped with the necessary hardware and software for receiving and sending CAN messages.

Devices using Ethernet POWERLINK, EtherCAT, or others can also implement the CANopen device profile. □

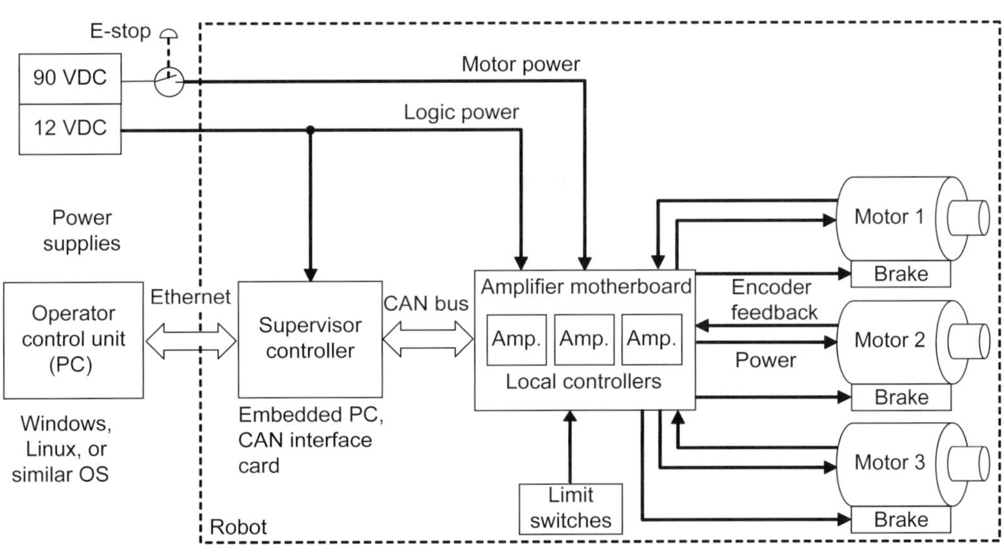

(a)

Figure 6.8 Networked control architecture for a three-axis robot: (a) hardware, (b) software.

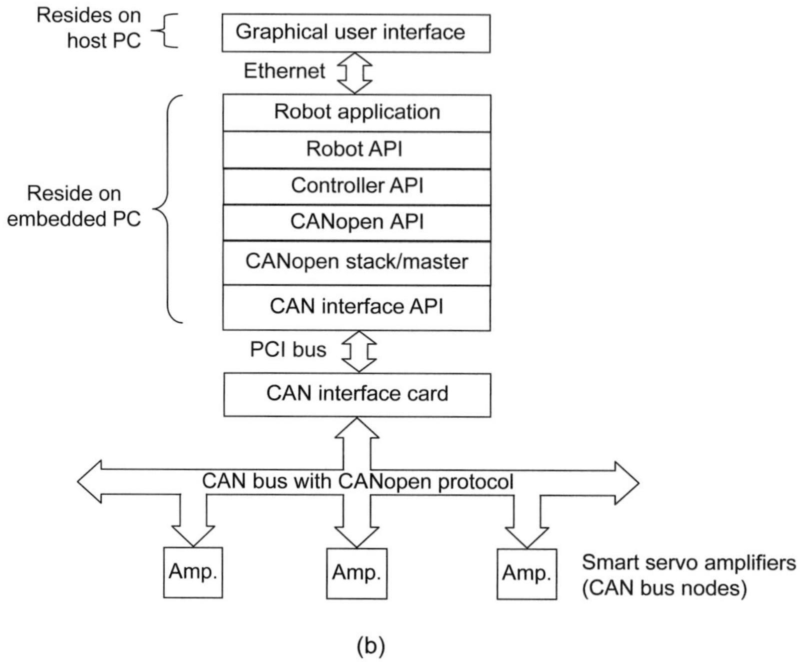

Figure 6.8 (cont.)

Example 6.4: coordinated control of a closed kinematic chain

The dual-arm vacuum robot in Example 4.8 has a cylindrical coordinate frame with five controllable axes of motion. The closed kinematic chains of both arms can be extended and retracted along their radial axes (R_1-axis for arm 1; R_2-axis for arm 2), and rotated about the Z-axis, along the common rotational Θ-axis. The Z-axis provides the vertical motion to both arms. Each arm is controlled by two independent revolute joints that apply torques to the respective inner links. The four revolute arm joints are concentric and have the same axis of rotation (the Z-axis, Figure 6.9). Arm extension and arm retraction along the R-axis is achieved by commanding the two joints per arm in opposite directions. Arm rotation along the Θ-axis is achieved by commanding both joints in the same direction while the robot base remains fixed in place. Accurate trajectory tracking of the end-effector requires precise coordination of the inner arm links, that is, the synchronization of both revolute joints. This is accomplished electronically using a feature that is sometimes called 'electronic gearing': one motor is defined as the master and the other the slave. The slave follows the master at a relative velocity that is specified by a programmable 'gear ratio' M:

$$\dot{\theta}_{slave} = M \cdot \dot{\theta}_{master}. \tag{6.60}$$

Encoder tracking is a technique for implementing electronic gearing: the master's position encoder is monitored, scaled by M, and then sent to the slave. For example, tracking the straight R-axis in Figure 6.9a requires a constant gear ratio of $M = -1$, while rotating along the Θ-axis at a constant R-position (tracking the circle in

Figure 6.9b) requires $M = 1$. A bus-based, decentralized controller would implement electronic gearing differently: the CANopen standard DS402 offers the interpolated position mode (IPM) for synchronized motion. This control mode is also called position/velocity/time (PVT), because the motor positions and velocities are synchronized with respect to time. When using the architecture in Figure 6.8 the supervisor controller first switches the two local controllers to the IPM mode and then transmits coarse position set points together with a time reference. The local controllers independently improve the trajectory resolution by interpolating the set points with respect to time, using a fixed cycle time. Then they calculate the necessary velocity and acceleration at all position set points. The supervisor controller then commands the execution of this arm motion and synchronizes both motors with a group synchronization signal. Without synchronization the local controllers would track the commanded trajectory independently using linear moves between two trajectory set points. □

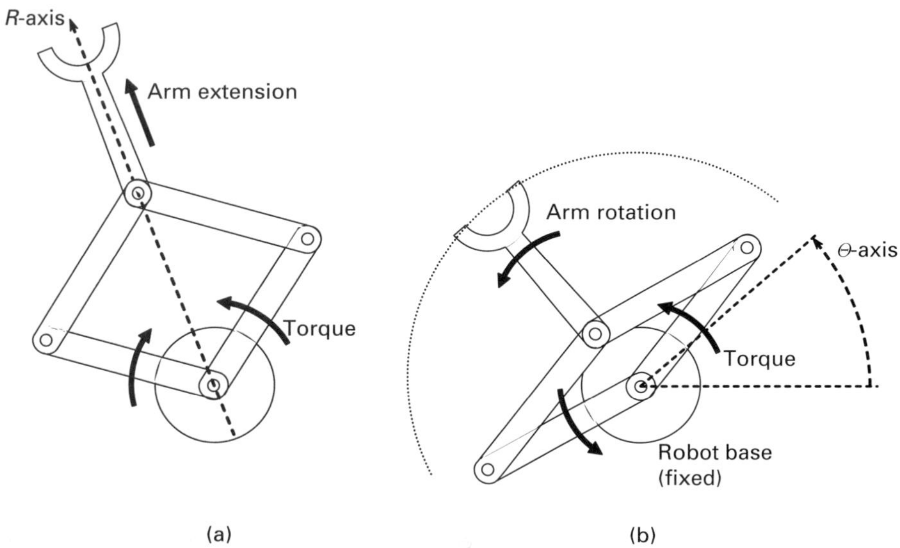

Figure 6.9 Controlling the closed-kinematic chain of a dual-arm robot.

6.4 Applicable and related standards

Several industry standards and guidelines apply directly or indirectly to the robots discussed in this chapter. The following list provides a selection. ANSI standards are published by the American National Standards Institute (ANSI). IEC standards are published by the International Electrotechnical Commission (IEC). CAN standards are published by CAN in Automation (CiA) and the International Organization for Standardization (ISO). RIA standards are published by the Robotic Industries

Association (RIA). SAE standards are published by Society of Automotive Engineers (SAE). SEMI standards are published by the Semiconductor Equipment and Materials International (SEMI). Contact information for these organizations is listed in Appendix B.

ANSI/RIA R15.02-1-1990, Human Engineering Design Criteria for Hand-Held Robot Control Pendants.

EN 50325-4:2002, CANopen – Industrial communication subsystem based on ISO 11898 (CAN) for controller–device interfaces.

IEC 61000-4-11 Ed. 2.0, Electromagnetic compatibility (EMC) – Part 4–11: Testing and measurement techniques – Voltage dips, short interruptions and voltage variations immunity tests.

ISO 11898-1:2003, Road vehicles – Controller area network (CAN) – Part 1: Data link layer and physical signalling.

SAE J1939/15, Reduced Physical Layer, 250K bits/s, Unshielded Twisted Pair (UTP).

SEMI F42-0600, Test Method for Semiconductor Processing Equipment Voltage Sag Immunity.

SEMI F47-0200, Specification for Semiconductor Processing Equipment Voltage Sag Immunity.

7 Test and characterization

This chapter presents procedures and relevant standards for testing and characterizing robots designed for electronics manufacturing in cleanroom environments. Several robot performance parameters are of interest: airborne particle contamination, surface particle contamination, positioning accuracy and repeatability, path accuracy and repeatability, vibration, and axis decoupling. The presented examples apply directly to substrate handling in semiconductor manufacturing and related industries. In the 1970s a typical semiconductor product yield, the number of semiconductor devices suitable for sale compared to the total number manufactured, was 10–15% in the USA. In Japan 60–90% was achieved through rigorous contamination control (Donovan, 2001). Today (2009) the product yield in modern semiconductor factories exceeds 90%, the result of contamination control equipment and procedures. Sections 7.1 and 7.2 present particle contamination tests.

7.1 Airborne particle contamination

Airborne particle contamination (APC) generated by substrate-handling robots can be a critical performance index, depending on the cleanliness requirements, and must be tested and characterized under normal robot operating conditions. The number of airborne particles is measured at selected locations in the immediate vicinity of the robot. For example, Sematech standard 92051107A-STD recommends measuring 'particles per wafer pass,' a performance criterion specified in SEMI standard E14. Substrate-handling robots should be tested and certified as subassemblies, and their contamination contribution specified as part of the total tool contamination allowance. An APC test procedure for individual robots is described below. Note that APC is different from airborne molecular contamination (AMC), a non-particle gaseous substance that can also be detrimental to a product or process. SEMI standard F21–95 defines four AMC categories: acids, bases, condensables, and dopants.

7.1.1 Clean mini-environments

A controlled mini-environment, a small cleanroom, is needed for particle testing. A typical mini-environment for robot testing is illustrated in Figure 7.1. The main building blocks are the enclosure, the fan-filter unit, the particle counter with probe and probe positioner, a robot mounting mechanism, and a connector to the external robot controller.

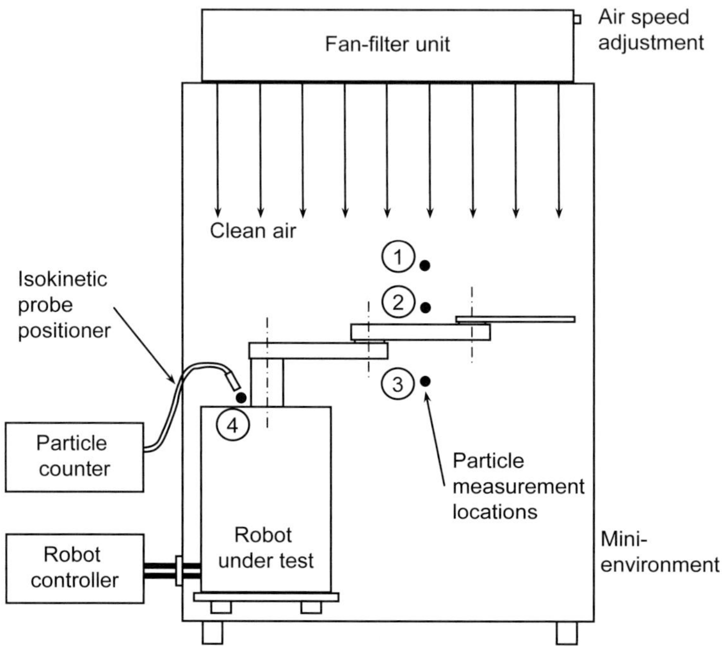

Figure 7.1 Mini-environment and robot for airborne particle testing.

A hermetic environment is not required, but a laminar flow of clean air per ISO Class 1 cleanliness is recommended. An extruded aluminum frame and Plexiglas® walls are suitable for the enclosure. Access to the interior is provided through a door sufficiently large for all expected robot models. An escape path for the laminar airflow is provided by perforated sheet metal at the bottom of the mini-environment. The airflow used during testing creates a positive gauge pressure, a slightly increased pressure in the mini-environment relative to the ambient atmosphere. Electrical connectors to the exterior controller and the probe positioner to the exterior particle counter are located near the bottom of the mini-environment in order not to impact the airflow.

The fan-filter unit at the top of the fixture provides laminar flow of clean, filtered air into the mini-environment. The fan-filter unit comprises the fan and a high-efficiency particulate air (HEPA®) filter. The standard requirement for HEPA® filters is a particle removal efficiency of at least 99.97% for all particles of 0.3 μm diameter or larger. Better HEPA® filters are also available; ultra-low particulate air (ULPA) of extreme cleanliness is achieved with filters that are 99.9995% efficient and remove particles of 0.12 μm or larger (IEST procedure IEST-RP-CC001.3; Schroth and Caesar, 2001; Winters, 2005). For ISO Class 1 cleanliness a 99.9995% filter efficiency for particle sizes of 0.12 μm or larger is recommended. The fan is equipped with an airspeed control. An air velocity of about $0.15\ m{\cdot}s^{-1}$ is typical. The particle counter is located outside the enclosure and is rated for ISO Class 1 cleanliness. It can detect airborne particles of 0.1 μm or larger at airflows up to $4.72 \cdot 10^{-4}\ m^3{\cdot}s^{-1}$ (1 cubic foot per minute). The flexible probe positioner allows access to measurement locations 1 to 4 as specified in Figure 7.1.

7.1.2 System operation

Cleaning and flushing

The mini-environment and robot must be cleaned before particle tests are performed. A typical cleaning procedure includes manually wiping the interior of the mini-environment with cleanroom-compatible, lint-free wipes and an appropriate solvent, for example isopropyl alcohol. Next the interior surfaces are cleaned with filtered, pressurized air, from top to bottom. The air escapes through the perforated floor and the open door. Then the door is closed and the fan-filter unit is turned on. The laminar flow of clean air flushes the interior and evacuates airborne particles through the perforated floor. After about 10 minutes the cleanliness of the mini-environment is measured with the particle counter. If no particles below a specified size, for example 0.1 μm, are detected during a measurement period of about 3 minutes the interior is considered clean. The mini-environment must be flushed again each time the door is opened.

Laminar airflow

Top-to-bottom laminar airflow and a positive gauge pressure inside the mini-environment permit the desired cleanliness: laminar airflow flushes particles away from the test area and through the exit holes in the perforated floor. A positive gauge pressure prevents contaminated ambient air from entering the mini-environment. However, increased airspeed can also create turbulent airflow, with vortices that can elevate particles back to the substrate. The air speed should be adjusted after robot installation such that no turbulent airflow is created. The air velocity and direction should be measured at several locations in the mini-environment. Some particle counters have an 'airflow measurement' mode for this purpose. Significant variations in velocity indicate turbulent airflow, in which case the air velocity must be reduced, while maintaining a positive gauge pressure. Typical air flow velocities are $0.3 - 0.5\ \text{m}\cdot\text{s}^{-1}$.

Probe positions and particle measurements

The particle counter probe should be close to the particle measurement locations, but without interfering with the robot's motion sequence. Particle tests are performed for at least 10 minutes at each specified location. Results are listed by particle size. All robots under test should perform the same motion sequence for each test in order to provide comparable results. Measurement locations depend on the robot type and model and on the motion sequence. Suggested locations for SCARA-type substrate-handling robots are shown in Figure 7.1:

- Position 1: 0.1 m above the end-effector, while the robot extends and retracts the arm (no vertical motion)
- Position 2: 0.01 m above the end-effector, while the robot extends and retracts the arm (no vertical motion)
- Position 3: 0.1 m below the end-effector, while the robot extends and retracts the arm (no vertical motion)

- Position 4: 0.01 m from the small gap between the vertical axis of motion and the top flange, while the robot extends and retracts the Z-axis (see Figure 3.5).

It can take several hours to test one robot, including one hour for preparations. However, not all manufactured robots must be tested. It is usually sufficient to perform particle tests on a relevant sample size for a particular robot model.

7.2 Surface particle contamination

This section presents methods for measuring substrate surface contamination generated by substrate-handling robots. Particles can be added by a robot to one or both surfaces of a substrate during handling. For example, a robot end-effector or aligner chuck can cause backside surface contamination.

7.2.1 Overview

Surface contaminants can be divided into three types: particle, atomic, and organic, including aerosols. Equipment contact and chemicals account for almost 90% of surface defects (Gao, 2004). The mechanical interface between the robot and the substrate is the end-effector. Typical end-effector parts and materials are anodized aluminum, PEEK pads, and cleanroom-compatible O-rings. Wafer handling is a source of systematic (as opposed to random) backside defects, through either damage from scratches or particle contamination.

Sematech standard 92051107A-STD specifies particles per wafer pass (PWP) as a quality measure for surface contamination from handling equipment. An example of a PWP specification is given in SEMI standard M52–0704, a guideline for surface inspection systems for silicon wafers. It states a desired particle contamination for the wafer backside of less than 0.001 PWP per cm^2 for a particle size of greater than 120 nm latex sphere equivalent (LSE), and greater than 90 nm LSE for the frontside. The desired particle contamination limit is the average number of particles added to a wafer surface during M passes, expressed as particles per unit area per pass (Tolliver, 1988). If N_I and N_F are the numbers of particles per unit area before and after the measurement, the number of particles added during one wafer pass is defined as:

$$N_{PWP} \leq \frac{N_F - N_I}{M} \quad , \text{unit} : \mathrm{m}^{-2} \text{ or } \mathrm{cm}^{-2}. \tag{7.1}$$

The result is rounded down if it is not an integer, thus the '≤' sign. Surface scanning is the detection method for surface particle contamination. Dark-field and bright-field illumination are the most common laser-based techniques.

The wafer backside is the unprocessed surface of a semiconductor wafer and may be used for handling. The backsides of 200 mm wafers are unpolished, those of 300 mm wafers polished. Yield-limiting backside defects are generally larger than frontside defects and can have several consequences. For example, backside particles can be pressed between the wafer and the end-effector and cause distortions of the front surface. In photolithography such distortions cause poor focus in small surface areas known as 'hotspots', while in chemical mechanical planarization (CMP) the undesired result is non-uniformities in removal rates and eventually variations in layer thickness (Carlson et al., 2007). The International Technology Roadmap for Semiconductors (ITRS) recommends defect detection limits of 40 nm (2007) and 30 nm (2010) for frontside defects, and 100 nm (2007) and 50 nm (2010) for backside defects (Saravanan et al., 2004). See also Figure 3.3. Another possible consequence of wafer handling is that backside particles can fall onto the processed front side of wafers stored below the contaminated wafer in the cassette slot.

Large, systematic backside defects from handling equipment can be classified using spatial pattern recognition (SPR) software, based on the size, shape, orientation, and position of defect patterns. Semiconductor manufacturers now catalogue the signatures of production tools and robotic systems to fully utilize the capabilities of SPR software, and often request drawings of mechanical interfaces that can leave recognizable defect patterns. The polished backside of 300 mm wafers facilitates surface scanning, whereas the rough backside of 200 mm wafers physically limits detection because particles in crevices and cracks are difficult to detect. It is evident that the contact area between the robot's mechanical interface (the end-effector) and the substrate must be minimized. Non-metallic contact is preferred in order to avoid metallic contamination: metals can produce particle contamination through micro-scratching of the substrate surface (Donovan, 2001).

7.2.2 Measurement system

A measurement system for surface contamination testing comprises a commercially available surface scanner, a robot with controller, and a PC with the test execution software installed. The objective is to measure the systematic surface contamination contributed by the robot's mechanical interface (end-effector). The surface scanner detects scattered laser light from particles on the wafer surface using a photodetector, and generates a particle map with particle sizes, based on the measured brightness of the scattered light.

The two main steps of a surface scan test are (a) cleaning of the mechanical interface and (b) measurement of the surface particle contamination deposited on an ultra-clean monitor substrate or wafer during a typical handling sequence. A monitor wafer is an ultra-clean wafer with fewer than 10 particles of 0.2 μm or larger. The entire test must be performed in an appropriate cleanroom environment, typically of ISO Class 2 cleanliness or better (see Table 2.1). The reliability of the test results increases with the sample size, that is, the number of monitor substrates used. In general two or three surface scan tests are sufficient per robot or end-effector. The test comprises the following steps:

Step 1. Clean and flush the mini-environment to remove existing contamination, using appropriate cleanroom procedures.

Step 2. Clean the robot's mechanical interface using a cleanroom-approved procedure.

Step 3. Scan the ultra-clean monitor substrate and document any existing surface contamination.

Step 4. Perform M initial wafer passes using the same monitor substrate and a typical handling sequence. No measurements are taken. This step is included in order to remove the remaining lose particles from the end-effector, because the goal is to count only particles generated by the robot's wafer handling.

Step 5. Scan the polished surface of the monitor substrate to determine N_I, the initial number of particles.

Step 6. Perform M final wafer passes using the monitor substrate and the same handling sequence.

Step 7. Scan the polished surface of the monitor substrate to determine N_F, the final number of particles.

Step 8. Determine N_{PWP} using Equation (7.1).

If more than one monitor substrate is used, the average N_{PWP} for all substrates should be calculated. See Example 7.1.

Example 7.1: backside surface scans of 300 mm wafers

This example presents the surface scans of two 300 mm monitor wafers (Figure 7.2). The backside contamination data was acquired with a surface scanning inspection system (KLA-Tencor SP2). The dots indicate detected particles, but do not indicate actual particle sizes. The wafer backside in Figure 7.2a shows systemic surface contamination from three end-effector O-rings and one aligner chuck in the center, and only two random particles. Figure 7.2b shows significant systemic particle contamination from a tool chuck or wafer cassette, an aligner chuck with concentric rings, end-effector O-rings, and numerous random particles. It can be seen that the second wafer was subjected to a manufacturing process, while the first was only handled and aligned to test the cleanliness of a robotic system. □

Example 7.2: wafer backside contamination

This example presents the results of two surface scans. The surface contamination of a robot end-effector for 300 mm wafers with a 2 mm edge exclusion zone was tested. The end-effector has three Kalrez® O-rings to minimize the contact area with the substrate. The sensitivity of the surface scanner was set to report particles with diameters larger than 0.20 μm, 0.48 μm, and 1.00 μm. One test was performed before and one test after a 48 hour 'burn-in,' during which the robot system handled clean test wafers. For the actual tests after the burn-in, two new, clean monitor wafers were used, one per test, and scanned before and after the wafer passes. With $M=2$ in Equation (7.1) the results are shown in Table 7.1.

The data shows that the end-effector cleanliness improved during the burn-in. It is concluded that during handling particles were removed from the O-rings by the monitor wafer. It also shows that the number of particles is roughly inversely proportional to their size. □

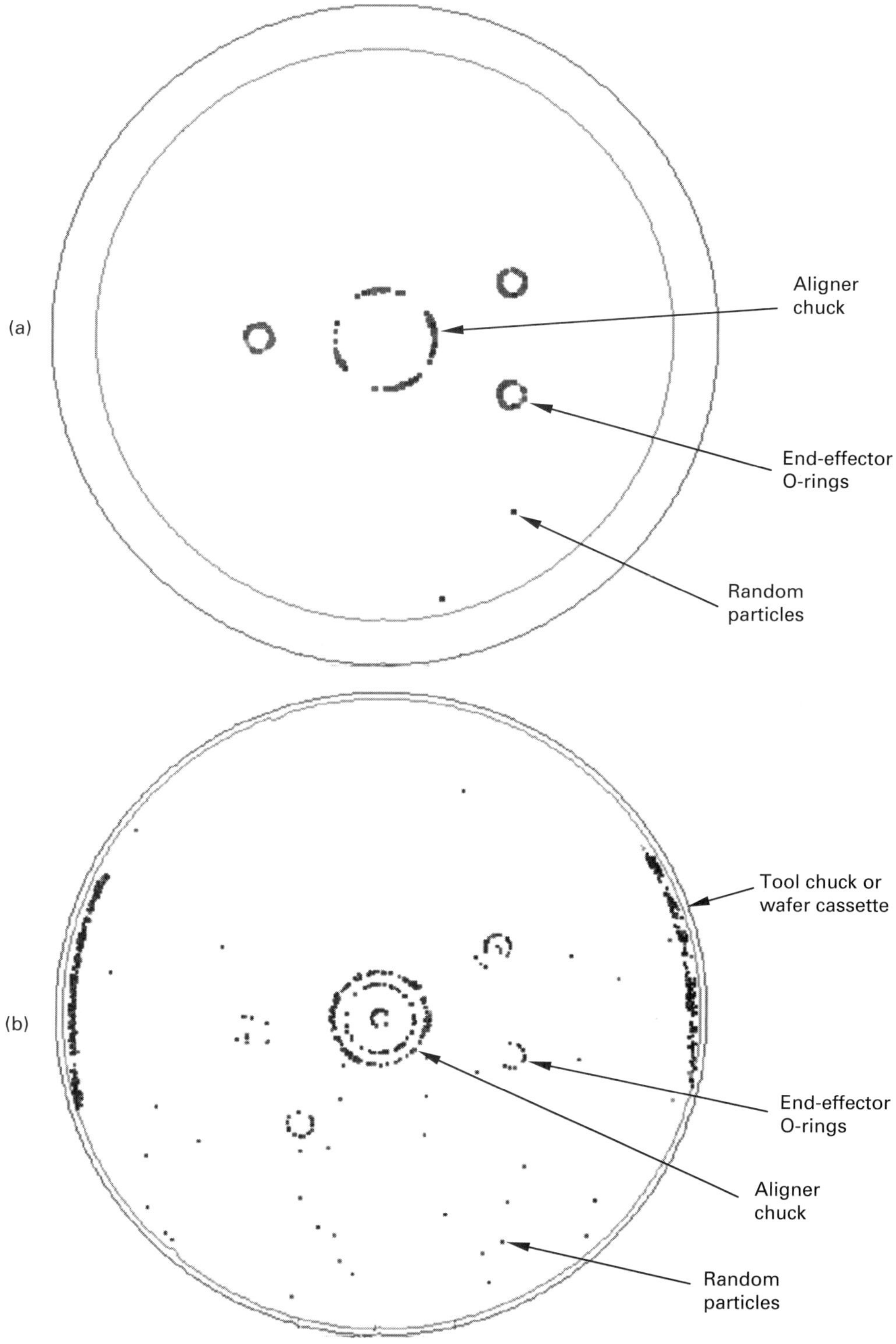

Figure 7.2 Backside particle contamination of two handled 300 mm wafers before (a) and after (b) subjected to a manufacturing process.

Table 7.1. Backside surface particle contamination, PWP by particle size ($M = 2$). Wafer area $A_W = \pi r^2 = \pi$ $(150\ \text{mm})^2 = 0.071\ \text{m}^2$.

Size limit	> 0.20 μm	> 0.48 μm	>1.00 μm
Before 48-hour burn-in:			
N_I	246	148	48
N_F	1044	713	282
N_{PWP}	399	282	117
After 48-hour burn-in:			
N_I	166	100	44
N_F	431	242	84
N_{PWP}	132	71	20

7.3 Positioning accuracy and repeatability

The characteristic variance in a robot's end-effector position as a function of the arm configuration (set of joint displacements) is the robot's 'signature.' Each robot has its own signature, due to its unique set of parts and manufacturing tolerances. The error contributions can be categorized as structural, kinematic, and dynamic (Shiakolas et al., 2002). A signature influences robot accuracy and repeatability. Accuracy can be described as the closeness in agreement between a test result, or the mean of several test results, and the commanded position. Repeatability can be described as the variation between nominal and measured dimensions (ISO vocabulary 1993):

- A robot's positioning accuracy is mostly affected by differences between the forward kinematics model implemented in the controller and the physical robot kinematics.
- Positioning repeatability is mostly affected by inconsistent parameters such as temperature and friction.

This section defines positioning accuracy and repeatability and presents methods for measuring and evaluating these performance indices. Path accuracy and repeatability are addressed in Section 7.4. Positioning accuracy and repeatability tests are recommended for all robots prior to shipment. Repeatability is more important than accuracy for certain applications, including substrate handling, where the fixed, commanded positions are first taught and revisited during operation. Two repeatability test methods are presented here, one for measurements in two dimensions (2D) and one in three dimensions (3D). A 2D repeatability test for substrate-handling robots in horizontal planes is discussed in Section 7.3.3. A repeatability test method for three dimensions is presented in Section 7.3.4.

7.3.1 Algorithm

ISO standard 9283 specifies tests for the measurement of several performance characteristics for manipulating industrial robots, including the cleanroom robots for electronics manufacturing discussed in this book. Positioning and orientation accuracy and repeatability are common performance characteristics, listed on the data sheets of virtually all industrial robots. ISO and ANSI/RIA standards require that this test covers a significant portion of the workspace at different commanded poses in the workspace, and that the results be averaged. See also the ANSI/RIA standard R15.05–2 and (Dagalakis, 1999).

Positioning and orientation accuracy

The pose (positioning and orientation) accuracy of a robot is defined as the deviation between a commanded pose and the mean of attained poses when the commanded pose is repeatedly approached from the same direction (ISO standard 9283). Both positioning accuracy a_P and orientation accuracy a_O determine the pose accuracy.

The positioning accuracy a_P of a robot is defined as the distance between the commanded pose $\boldsymbol{x}_{cmd}$ and the mean $\bar{\boldsymbol{x}}$ of N attained positions, with

$$\boldsymbol{x}_{cmd} = \begin{bmatrix} x_{cmd} \\ y_{cmd} \\ z_{cmd} \end{bmatrix}, \quad \bar{\boldsymbol{x}} = \begin{bmatrix} \bar{x} \\ \bar{y} \\ \bar{z} \end{bmatrix}. \tag{7.2}$$

The mean coordinates are given by

$$\bar{x} = \frac{1}{N}\sum_{i=1}^{N} x_i, \ \bar{y} = \frac{1}{N}\sum_{i=1}^{N} y_i, \ \bar{z} = \frac{1}{N}\sum_{i=1}^{N} z_i. \tag{7.3}$$

x_i, y_i, z_i are the position coordinates of the i-th measured pose. The positioning accuracy a_P is obtained using the Euclidean norm:

$$\begin{aligned} a_P &= \|\bar{\boldsymbol{x}} - \boldsymbol{x}_{cmd}\| \\ &= \sqrt{(\bar{x} - x_{cmd})^2 + (\bar{y} - y_{cmd})^2 + (\bar{z} - z_{cmd})^2}. \end{aligned} \tag{7.4}$$

A robot's orientation accuracy a_O is defined for each axis of motion. It is the difference between the orientation angles $\alpha, \ \beta, \ \gamma$ of the commanded pose, and the mean angles $\bar{\alpha}, \bar{\beta}, \bar{\gamma}$ of the N attained angles, for each axis. The orientation angle α is associated with the x-axis, β with the y-axis, and γ with the z-axis:

$$\begin{aligned} a_\alpha &= (\bar{\alpha} - \alpha_{cmd}) \\ a_\beta &= (\bar{\beta} - \beta_{cmd}) \\ a_\gamma &= (\bar{\gamma} - \gamma_{cmd}). \end{aligned} \tag{7.5}$$

The mean angles are given by

$$\bar{\alpha} = \frac{1}{N}\sum_{i=1}^{N} \alpha_i, \ \bar{\beta} = \frac{1}{N}\sum_{i=1}^{N} \beta_i, \ \bar{\gamma} = \frac{1}{N}\sum_{i=1}^{N} \gamma_i, \tag{7.6}$$

where $\alpha_i, \beta_i, \gamma_i$ are the orientation angles of the i-th measured pose.

It is convenient to set the commanded position to zero via a shift of coordinates:

$$\boldsymbol{x}_i^* = \boldsymbol{x}_i - \boldsymbol{x}_{cmd}, \quad i = 1, 2, \cdots, \ N, \tag{7.7}$$

and then compute the accuracy using the above algorithm.

Positioning and orientation repeatability

The pose repeatability is defined in terms of the agreement among the attained poses for N repeat visits to the same command pose from the same direction (ISO standard 9283). The SEMI standard E89 in addition recommends performing repeatability tests under 'repeatability conditions': the same operator uses the same equipment and the same procedure at consistent ambient conditions. The repeatability of a given commanded robot pose is determined by two parameters: positioning repeatability r_{P} and orientation repeatability r_{O}.

The positioning repeatability r_{P} is defined as the radius of the disc or sphere about the mean position $\bar{\boldsymbol{x}}$,

$$r_P = \bar{d} + 3\sigma_d. \tag{7.8}$$

See Figure 7.3. $\bar{d}$ is the mean of all distances between the mean position $\bar{\boldsymbol{x}}$ in Equation (7.2) and the N attained positions in the point cloud:

$$\begin{aligned} \bar{d} &= \frac{1}{N}\sum_{i=1}^{N} d_i \\ &= \frac{1}{N}\sum_{i=1}^{N}\left(\sqrt{(x_i - \bar{x})^2 + (y_i - \bar{y})^2 + (z_i - \bar{z})^2}\right). \end{aligned} \tag{7.9}$$

The mean position coordinates $\bar{x}, \bar{y}, \bar{z}$ are defined in Equation (7.3). σ_d is the standard deviation of all measured $d_i, i = 1, 2, \cdots, \ N$:

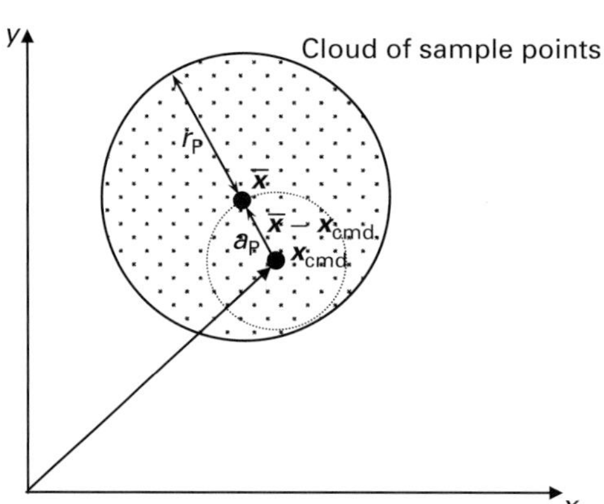

Figure 7.3 Definitions of positioning accuracy a_{P} and repeatability r_{P}.

$$\sigma_d = \sqrt{\frac{\sum_{i=1}^{N}(d_i - \bar{d})^2}{N-1}}. \tag{7.10}$$

Gaussian probability theory states that for normally distributed data the range $3\sigma_d$ contains 99.7% of all sample points (Papoulis, 1991). If the data deviate from a normal distribution the percentage of data points within the $3\sigma_d$ disc will deviate from the theoretical 99.7%.

The orientation repeatability r_O is expressed in terms of individual repeatabilities $r_\alpha, r_\beta, r_\gamma$, the 3σ distributions of the orientation angles α, β, γ, respectively:

$$\begin{aligned} r_\alpha &= \pm 3\sigma_\alpha = \pm 3 \cdot \sqrt{\frac{\sum_{i=1}^{N}(\alpha_i - \bar{\alpha})^2}{N-1}} \\ r_\beta &= \pm 3\sigma_\beta = \pm 3 \cdot \sqrt{\frac{\sum_{i=1}^{N}(\beta_i - \bar{\beta})^2}{N-1}} \\ r_\gamma &= \pm 3\sigma_\gamma = \pm 3 \cdot \sqrt{\frac{\sum_{i=1}^{N}(\gamma_i - \bar{\gamma})^2}{N-1}}. \end{aligned} \tag{7.11}$$

Overall positioning performance

Although accuracy and repeatability are specified separately, both are needed to describe positioning performance. Figure 7.4 illustrates the difference between accuracy and

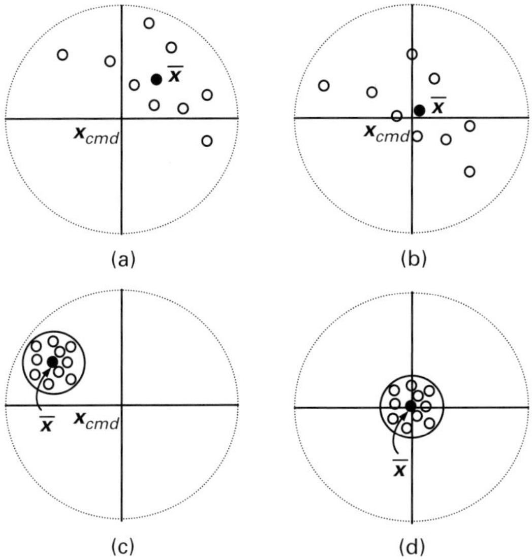

Figure 7.4 Illustration of positioning accuracy and repeatability.

repeatability. The commanded position is placed at $\boldsymbol{x}_{cmd} = \mathbf{0}$. The smaller circle in the figure indicates the positioning repeatability r_P.

- Figure (a): poor a_P, poor r_P ($\bar{\boldsymbol{x}}$ deviates from $\boldsymbol{x}_{cmd}$ and the points are scattered)
- Figure (b): good a_P, poor r_P ($\bar{\boldsymbol{x}}$ is close to $\boldsymbol{x}_{cmd}$ and the points are scattered)
- Figure (c): poor a_P, good r_P ($\bar{\boldsymbol{x}}$ deviates from $\boldsymbol{x}_{cmd}$ and the points are clustered)
- Figure (d): good a_P, good r_P ($\bar{\boldsymbol{x}}$ is close to $\boldsymbol{x}_{cmd}$ and the points are clustered).

Discussion

Given limited test data and the statistical measures a_P and r_P, what should be the assumed worst-case positioning performance? The following is a heuristic method suggested for test data that covers only a limited portion of the workspace. The worst-case performance is estimated based on the limited data. A non-zero and consistent accuracy a_P is usually the result of an inaccurate kinematics model, while a non-zero r_P is typically a result of noise in sensor and control signals, temperature variations, or tolerances in the electro-mechanical system (Conrad et al., 2000).

- It is assumed that for the robot under test $\bar{\boldsymbol{x}}$ is consistent in a specified portion of the workspace. Given constant a_P and r_P, this is illustrated in Figure 7.5 with the small circle of radius a_P about $\boldsymbol{x}_{cmd}$.
- a_P and r_P are measured in the specified portion of the workspace. It is assumed that in different workspace portions the 'directions' of a_P and r_P change, depending on the robot's parameters and tolerances. The worst-case performance is then the larger circle about $\boldsymbol{x}_{cmd}$ with constant radius $a_P + r_P$ (Figure 7.5).

A 'cross section' of the probability values in Figure 7.5 is shown in Figure 7.6. The Gaussian 'bell' shows the probability of sample point locations with respect to $\boldsymbol{x}_{cmd}$ and $\bar{\boldsymbol{x}}$.

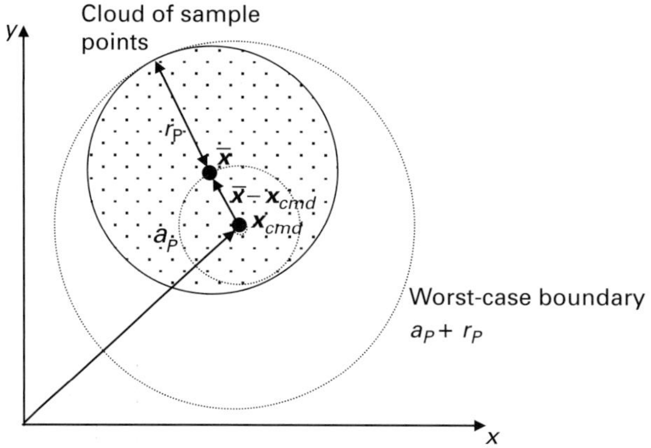

Figure 7.5 Assumed worst-case positioning performance based on a_P and r_P.

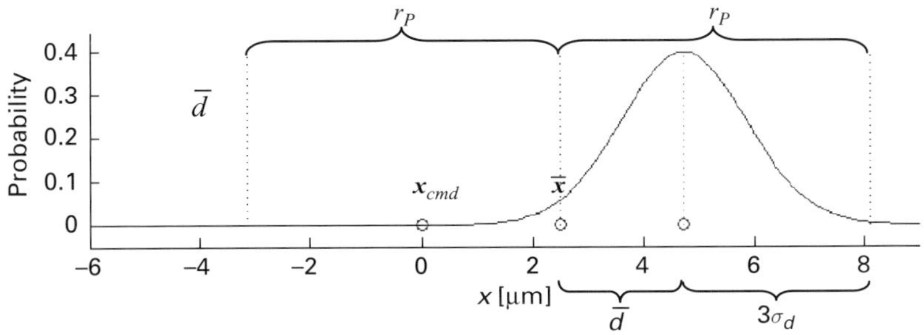

Figure 7.6 Probability 'cross section' of the disc in Figure 7.5.

Position stabilization time

The position stabilization time is a measure of robot performance that quantifies how quickly robot vibration settles at the attained pose (ISO standard 9283). The end-effector position must be continuously measured at a sufficient sampling rate as the robot approaches the commanded pose. The position stabilization time is the elapsed time from first entering a specified tolerance zone until the end-effector remains within that zone. The tolerance zone can be defined by the user or taken from the robot manufacturer's data sheet. The robot should be tested with the same motion used for measuring accuracy and repeatability, and with the same test load and test velocities. This procedure is repeated three times and the mean value of the elapsed time is reported.

7.3.2 Performing the test

The above accuracy and repeatability tests require a measurement system, a mini-environment, and a well defined test procedure. The tests are valid only under the environmental and normal operating conditions specified by the robot manufacturer (ISO standard 9283).

Robot

The test robot must be completely assembled and fully operational. All necessary calibration, leveling, and alignment procedures, as well as functional tests, must have been completed. The robot is mounted according to the manufacturer's recommendations. If specified by the manufacturer, an appropriate number of warm-up cycles are performed. The warm-up period varies, although a minimum of 10 minutes is suggested.

Measurement system

The measurement system comprises hardware and software:

- Sensors: contact-free measurements are recommended in order to avoid influencing the test results.
- Data acquisition system: measures and logs the physical entity in question (here: end-effector or substrate position) and interfaces with a computer.

- Data analysis software: transforms data into useful information using statistics and other algorithms.
- Test execution program: automates the test by executing and coordinating all activities for the required number of test cycles.

Reproducibility of the measurement system, that is, the repeatability of test data under constant conditions, can be tested using elaborate procedures defined, for example, by the SEMI standard E89-1104.

Mini-environment and operating conditions

The mini-environment (or cleanroom laboratory) where the test is to be performed must provide realistic environmental conditions. These include cleanliness, temperature, relative humidity, and electromagnetic and electrostatic fields. The ambient temperature in the mini-environment should be maintained at 20±2 °C (ISO standard 9283). Exceptions must be stated and explained in the test report. The robot and the sensors should be located in the test environment long enough to reach a thermally stable condition.

Procedure

Measuring accuracy and repeatability involves several steps:

- Preparation of the robot: it must be completely assembled, fully operational, mounted as specified, and warmed up.
- Preparation of the mini-environment: maintain normal operations and environmental conditions; place sensors in the mini-environment for several hours prior to the test.
- Calibration of the measurement system: align the coordinate systems of robot and measurement system and initialize all test variables.
- Data acquisition: perform N robot test cycles (repeat visits) from the initial pose to the commanded pose at the sensors.
- Data analysis: compute positioning accuracy and repeatability from the logged test data.

Two examples of accuracy and repeatability measurement systems are presented below. The position and orientation of the substrate (not the end-effector) are measured in order to quantify robot performance and possible substrate slippage on the end-effector. A test wafer with reference markers is used.

7.3.3 Vision-based repeatability test in two dimensions

Substrate-handling applications in electronics manufacturing typically require good repeatability for horizontal motion, but less stringent requirements exist for the vertical axis of motion. The vision-based measurement system presented below is designed for such planar repeatability tests.

Repeatability measurement system

The repeatability measurement system with its subsystems is illustrated in Figure 7.7. The position sensors are two cameras at fixed locations. The cameras establish the

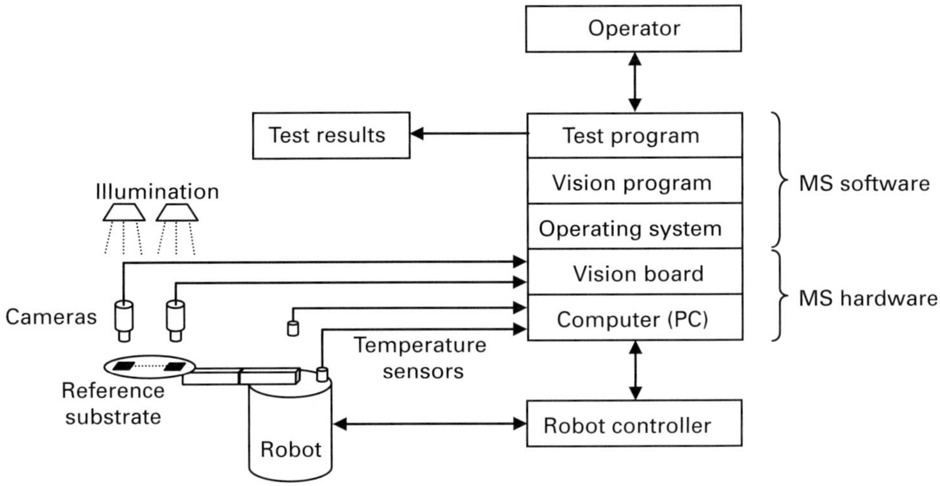

Figure 7.7 Diagram of 2D accuracy and repeatability measurement system (MS).

measurement system's reference coordinate frame. Each camera acquires a reference pattern attached to a test substrate (here: a wafer). A vision program processes the two digital images and determines their position and orientation in the measurement system reference frame. The test program automates the entire repeatability test, including the coordination of robot motion, data acquisition, and data analysis. The human operator only starts and monitors the test. The temperature sensors measure the ambient temperature and interior robot temperature in order to detect any correlation between temperature variations and the robot's accuracy and repeatability performance.

The vision system comprises two digital cameras, a vision board (a microprocessor with appropriate digital I/O), suitable illumination sources, and a vision program that processes the acquired images and detects the relevant features. A reference pattern and its use are illustrated in Figure 7.8. The pattern is firmly attached to the test wafer and aligned with the cameras, and as such with the reference frame O-xy. The commanded position and orientation $\boldsymbol{x}_{cmd}$ are shown in the lower part of Figure 7.8. The deviation from $\boldsymbol{x}_{cmd}$ at the i-th repeat visit to $\boldsymbol{x}_{cmd}$ is illustrated in the upper part of Figure 7.8 with exaggerated positioning error e_P and wafer orientation error e_O.

The test procedure involves several steps:

Step 1. Kinematic robot calibration ('teaching'): move the robot to the commanded pose $\boldsymbol{x}_{cmd}$ such that the two test patterns are within the cameras' fields of view. Specify the midpoint of the two cameras (the origin O) as the commanded robot pose $\boldsymbol{x}_{cmd}$.

Step 2. Calibration of the measurement system: align the measurement system's reference frame O-xy with the end-effector coordinate frame. Teach the vision program both reference pattern positions at the commanded pose $\boldsymbol{x}_{cmd}$.

Step 3. Performing the test: move the robot N times from the initial to the commanded end-effector position. Measure the pattern positions each time and compute the

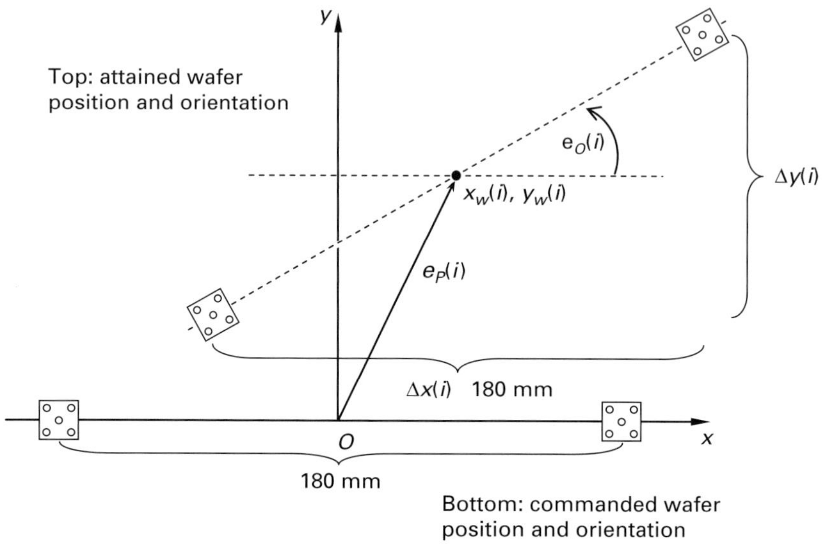

Figure 7.8 Use of a reference pattern for measuring (positioning) accuracy and repeatability.

attained wafer position at the wafer center, and the wafer orientation, using Equations (7.12) and (7.13) below.

Step 4. Data analysis: after N test cycles compute position and orientation repeatability using Equations (7.8) and (7.11).

This test can also be used to measure substrate slippage on the end-effector during motion. First the robot is tested with the substrate firmly attached to the end-effector. For the second, identical test the typical gripping mechanism, for example vacuum suction, is used to hold the substrate. Any difference in the results is due to substrate slippage.

The robot motion used for the test should be typical of normal operation, for example unloading a wafer from a cassette. The test should be repeated with different initial positions. Either the average repeatability performance (ISO standard 9283) or the worst case is reported. ISO 9283 recommends $N = 30$ repeat visits per test.

Computing position and orientation errors

The substrate position $\boldsymbol{x}(i)$, measured at the i-th repeat visit to $\boldsymbol{x}_{cmd}$, is the midpoint between the measured positions at camera 1, $\boldsymbol{x}_{C1}(i) = [\,x_1(i) \quad y_1(i)\,]^T$ and camera 2, $\boldsymbol{x}_{C2}(i) = [\,x_2(i) \quad y_2(i)\,]^T$. The midpoint is

$$\boldsymbol{x}(i) = \frac{1}{2}(\boldsymbol{x}_{C1}(i) + \boldsymbol{x}_{C2}(i)), \; i = 1, \cdots, \; N. \tag{7.12}$$

The position repeatability (3σ) is then computed using Equation (7.8).

The substrate orientation (or angle) $\theta(i), \; i = 1, 2, \cdots, \; N$, is also obtained from $\boldsymbol{x}_{C1}(i)$ and $\boldsymbol{x}_{C2}(i)$. The commanded wafer angle is $\theta_{cmd} = 0$. The wafer angle errors are typically smaller than $0.01°$, so the small-angle theorem applies, $\theta \approx \sin(\theta)$:

$$\theta(i) = \tan^{-1}\left(\frac{\Delta x(i)}{\Delta y(i)}\right)\text{rad} = \tan^{-1}\left(\frac{y_1(i) - y_2(i)}{0.18\text{m}}\right) \cdot \frac{180}{\pi}\text{deg.} \tag{7.13}$$

The distance between the patterns is 180 mm. The orientation repeatability (3σ) is then obtained using Equation (7.11). Note that the angles $\theta(i)$ are independent from $\boldsymbol{x}(i)$. The substrate's angle repeatability is of special interest if an angle aligner is used within a process tool on the manufacturing floor.

Measurement resolution

The measurement system's resolution depends on the vision system. It is recommended that the complete camera field of view be used for a test pattern to achieve the maximum gauge resolution. A camera resolution of 512×480 pixels and a test pattern size of 10.0×10.0 mm results in a horizontal resolution of 10/512 = 0.0195 mm and a vertical resolution of 10/480 = 0.0208 mm. The two test patterns are 180 mm apart. The angular resolution is therefore $\tan^{-1}(0.0208\ \text{mm}/180\ \text{mm}) = 0.0066°$. If the vision system can accurately locate a pattern with a (technically feasible) 0.25-pixel resolution, resolutions of 0.0049 mm and 0.0052 mm can be achieved. The angular resolution would be $\tan^{-1}(0.0052/180) = 0.0017\,°$.

The gauge repeatability of the measurement system was improved with the following adjustments:

- The distance between the two illuminating fluorescent lamps and the cameras was increased from 0.05 m to 0.30 m (Figure 7.7). It is assumed that the temperature in the immediate proximity of the lamps affected the measurements.
- The cameras were turned on for two hours prior to the test.
- The environment was protected from physical disturbances that could cause any vibration of the measurement system.
- The ambient temperature was maintained at a constant level within the facility's accuracy (2 °C).

The following examples illustrate the algorithms used by the vision-based measurement system.

Example 7.3: 2D repeatability and accuracy

This numerical example illustrates how the measurement system in Section 7.3.3 computes positioning accuracy a_p and repeatability r_p. The substrate center positions $\boldsymbol{x}(i)$ and angles $\theta(i)$) are computed using the following series of pattern positions (x_{C1}, x_{C2}) measured with cameras 1 and 2. All measurements and results are in μm or degrees:

$$\boldsymbol{x}_{C1}(i) = \begin{bmatrix} x_{C1}(i) \\ y_{C1}(i) \end{bmatrix} = \left(\begin{bmatrix} 2 \\ 3 \end{bmatrix}, \begin{bmatrix} 4 \\ -10 \end{bmatrix}, \begin{bmatrix} -3 \\ -3 \end{bmatrix}, \begin{bmatrix} 1 \\ 10 \end{bmatrix}, \begin{bmatrix} 0 \\ 2 \end{bmatrix}\right), \quad i = 1, \cdots, 5$$
$$\boldsymbol{x}_{C2}(i) = \begin{bmatrix} x_{C2}(i) \\ y_{C2}(i) \end{bmatrix} = \left(\begin{bmatrix} 2 \\ -1 \end{bmatrix}, \begin{bmatrix} 4 \\ 6 \end{bmatrix}, \begin{bmatrix} -3 \\ 9 \end{bmatrix}, \begin{bmatrix} 1 \\ -6 \end{bmatrix}, \begin{bmatrix} 0 \\ -4 \end{bmatrix}\right), \quad i = 1, \cdots, 5. \tag{7.14}$$

The substrate center position errors in the measurement system's reference frame are obtained using Equation (7.12):

$$\boldsymbol{x}(i) = \left(\begin{bmatrix} 2 \\ 1 \end{bmatrix}, \begin{bmatrix} 4 \\ -2 \end{bmatrix}, \begin{bmatrix} -3 \\ 3 \end{bmatrix}, \begin{bmatrix} 1 \\ 2 \end{bmatrix}, \begin{bmatrix} 0 \\ -1 \end{bmatrix} \right)$$
$$\bar{\boldsymbol{x}} = \begin{bmatrix} 0.8 \\ 0.6 \end{bmatrix}. \tag{7.15}$$

The angle errors are obtained using Equation (7.13):

$$\theta(i) = (0.0013°, -0.0051°, -0.0038°, 0.0051°, 0.0019°). \tag{7.16}$$

The positioning accuracy a_P is computed using Equation (7.4):

$$a_P = \left\| \begin{bmatrix} 0.8 \\ 0.6 \end{bmatrix} - \begin{bmatrix} 0 \\ 0 \end{bmatrix} \right\| = 1. \tag{7.17}$$

The positioning repeatability r_P is obtained from Equations (7.8), (7.9), and (7.10):

$$\bar{d} = 2.6$$
$$\sigma_d = \sqrt{\frac{\sum_{i=1}^{N} (d_i - \bar{d})^2}{N-1}} = \sqrt{\frac{0.9987}{4}} = 0.4997 \tag{7.18}$$
$$r_P = \bar{d} + 3\sigma_d = 7.3. \square$$

Example 7.4: 2D vision-based repeatability test

This example presents the measurement results, including positioning accuracy a_P and repeatability r_P, for a tested wafer-handling robot. Figure 7.9 shows two-dimensional measurement points acquired by the vision-based measurement system of Section 7.3.3. The positioning accuracy and repeatability results were obtained from 3535 sample points using Equations (7.8), (7.9), and (7.10), and are summarized in Table 7.2. Note that only 42 sample points (black dots) are shown in Figure 7.9. This is a consequence of the 1 μm resolution of the measurement system. The commanded robot pose $\boldsymbol{x}_{cmd} = \mathbf{0}$ is marked with an X. The mean position $\bar{\boldsymbol{x}}$ is marked with an O. This is the center of the repeatability circle with radius r_P in Equation (7.8). Only 21 position samples (0.59%) are outside the repeatability circle, which is close to the theoretical value of 11 samples, or 0.30%. □

7.3.4 Laser-based repeatability test in three dimensions

This section presents a laser-based system that measures positioning accuracy and repeatability in three dimensions, including the vertical deflection of a robot arm. The

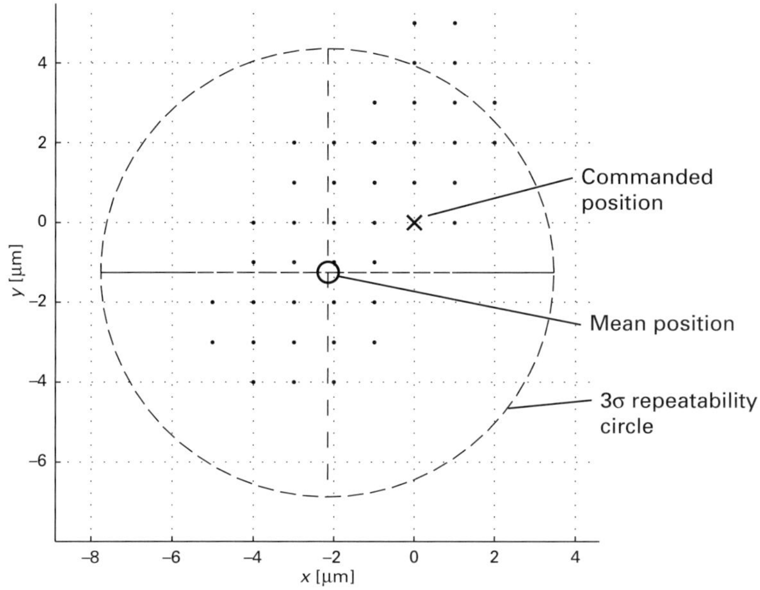

Figure 7.9. Accuracy and repeatability results for Example 7.4.

Table 7.2. Repeatability and accuracy results for Example 7.4.

Parameter	Value
Number of measurements	$N = 3\,535$
Commanded pose	$\boldsymbol{x}_{cmd} = (0,0)$
Mean pose	$\bar{\boldsymbol{x}}_w = (-0.002\,1, -0.001\,3)$
Accuracy	$a_P = 0.002\,5$
Mean distance	$\bar{d} = 0.002\,2$
Standard deviation	$\sigma_d = 0.001\,1$
Repeatability	$r_P = 0.005\,6$
$3\sigma_d$ outliers	$N_{out} = 21$ (0.59%)

algorithms in Section 7.3.1 for computing positioning accuracy and repeatability apply. The end-effector or substrate orientation cannot be determined because only one position is measured per test cycle.

Measurement system

The measurement system utilizes three commercially available laser proximity sensors with a resolution of ±1.0 μm, and a reflective target (a 12 mm, non-glossy white cube), and an appropriate fine-adjustment stage. The data acquisition system measures and logs the substrate or end-effector positions and interfaces with a computer. The data analysis software converts the raw data into physically meaningful data using the algorithms in Section 7.3.1. The test execution program automates the test.

Many laser proximity sensors utilize the triangulation principle to measure distances to a target. A laser source emits a beam that is reflected by the target and detected by a

Example 7.5: 3D laser-based repeatability test

This example presents the calculation of positioning accuracy and repeatability for an atmospheric wafer-handling robot that was tested in three dimensions using the laser-based measurement system. The robot has a SCARA-type arm with two links (length: 179 mm) and a 'blade-type' end-effector with vacuum suction (length: 190 mm). The commanded robot pose was an almost fully extended arm, close to the boundary of the robot's cylindrical workspace. Figure 7.11 shows the data point cloud projected onto the *x*/*y*, *x*/*z*, and *y*/*z* planes, together with the repeatability boundaries and the commanded position ('X'). The mean position was shifted to the origin at the center of the crosshairs. The repeatability boundaries, one per plane, are also shown in the 3D view with a 37.5° azimuth and a 30° elevation. The commanded position is close to the r_P sphere. There are only two outliers near the commanded position, although they are not recognizable in the figure. Note the S-shaped structure of the data, which may be due to several factors, including a changing temperature that affects the robot kinematics over time.

The positioning accuracy a_P is obtained using Equation (7.4):

$$a_P = \|\bar{\boldsymbol{x}} - \boldsymbol{x}_{cmd}\| = \left\| \begin{bmatrix} 0.0275 \\ 0.0292 \\ -0.0332 \end{bmatrix} - \begin{bmatrix} 0 \\ 0 \\ 0 \end{bmatrix} \right\| \text{mm} = 0.0520\,\text{mm}. \tag{7.19}$$

The positioning repeatability r_P is obtained using Equation (7.8):

$$r_P = \bar{d} + 3\sigma_d = (0.0213 + 3 \cdot 0.0103)\ \text{mm} = 0.0523\ \text{mm}. \tag{7.20}$$

The results are summarized in Table 7.3. □

Table 7.3. Laser-based 3D repeatability test (Example 7.5).

Parameter	Value
Number of measurements	$N = 1\,000$
Commanded pose	$\boldsymbol{x}_{cmd} = (0,0,0)$
Mean pose (mm)	$\bar{\boldsymbol{x}} = (0.027\,5, 0.029\,2, -0.033\,2)$
Accuracy (mm)	$a_P = 0.052\,0$
Mean distance (mm)	$\bar{d} = 0.021\,3$
Standard deviation (mm)	$\sigma_d = 0.010\,3$
Repeatability (mm)	$r_P = 0.052\,3$
$3\sigma_d$ outliers	$N_{out} = 2$ (0.2%)

sensor. The laser position on the sensor is compared with a reference position, which allows the distance to the target to be inferred. Figure 7.10 shows three sensors, mounted in line with the axes of the measurement system's Cartesian reference frame *O-xyz*. The three laser beams are reflected by the target cube whose *x*, *y*, and *z* positions are measured. (In practice the origin *O* is at the target.)

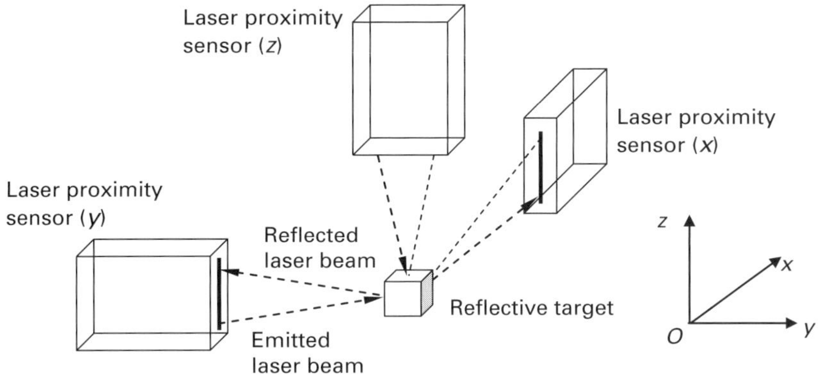

Figure 7.10 Laser-based position measurements in three dimensions.

A test procedure similar to the one described in Section 7.3.3 applies. Normal operation and environmental conditions are required prior to and during the test. The robot must be securely mounted to the measurement system to avoid any relative motion during the test. The reflective target is attached to the end-effector or the substrate. The end-effector is moved to the reference position within range of the sensors. The laser sensors are then positioned at the reference distances from the target. A fine-adjustment stage allows for the individual calibration of each sensor (Figure 7.10).

The repeatability test is executed for N test cycles, during each of which the robot is commanded to move from the initial to the reference end-effector position within range of the sensors. The minimal number of test cycles recommended by ISO standard 9283 is $N = 30$. The robot position is measured at each cycle. The logged position data is analyzed and the positioning accuracy and repeatability determined and saved. The algorithms in Section 7.3.1 can be applied directly, because the $\boldsymbol{x}(i),\ i = 1, \ldots, N$ positions are measured and available.

7.4 Path accuracy and repeatability

Path accuracy and repeatability describes a robot's ability to repeatedly move its end-effector along the same commanded path in the same direction. It is defined as the maximum path deviation along the commanded path in terms of positioning and orientation. This section discusses the motivation for measuring the path (positioning) accuracy and repeatability in electronics manufacturing, and the definitions and algorithms in ISO standard 9283. The algorithms apply to any path shape. A laser-based measurement system for testing path positioning accuracy a_T and repeatability r_T is then described.

A motivation for measuring a_T and r_T of substrate-handling robots arises from the tight geometry of manufacturing tools and substrate carriers. For example, a small and appropriate value of a_T of a straight path in a horizontal wafer plane is required for picking and placing wafers in tight substrate carriers. Any contact between substrates, or between a substrate and the cassette, during robot motion can generate particles or

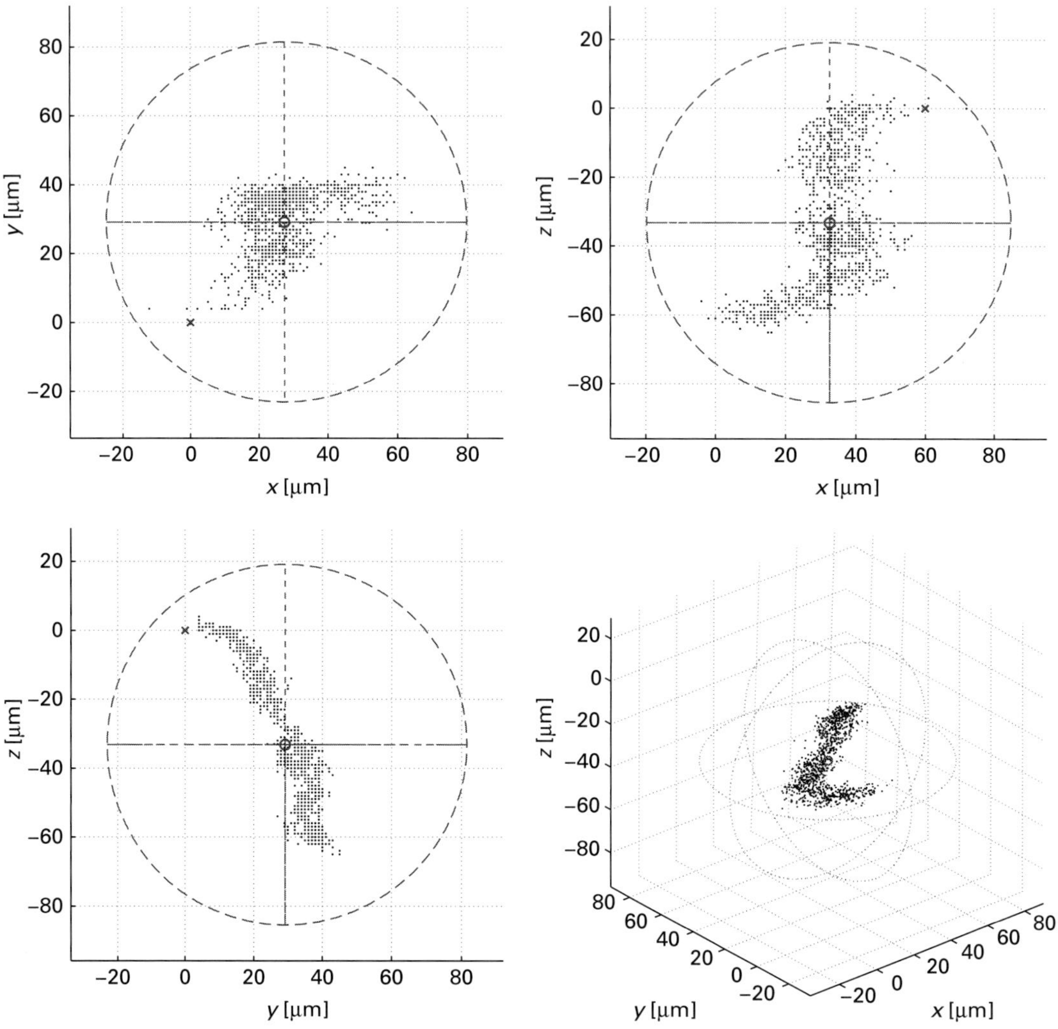

Figure 7.11 3D repeatability data and results for Example 7.5.

damage the product. Other examples are tight load locks or doors through which wafers are transferred. This is addressed in SEMI standard E21-94 for vacuum cluster tools, where substrates are handled in horizontal wafer transport planes with a vertical path accuracy of ±0.5 mm.

7.4.1 Definitions and algorithms

The definitions and algorithms presented here are adopted from ISO standard 9283. a_{T} and r_{T} for three-dimensional paths are related to the positioning accuracy a_{P} and repeatability r_{P} for two-dimensional datasets. See Equations (7.4) and (7.8). The measurements for a_{T} and r_{T} are taken in planes perpendicular to the commanded path ξ_{cmd}.

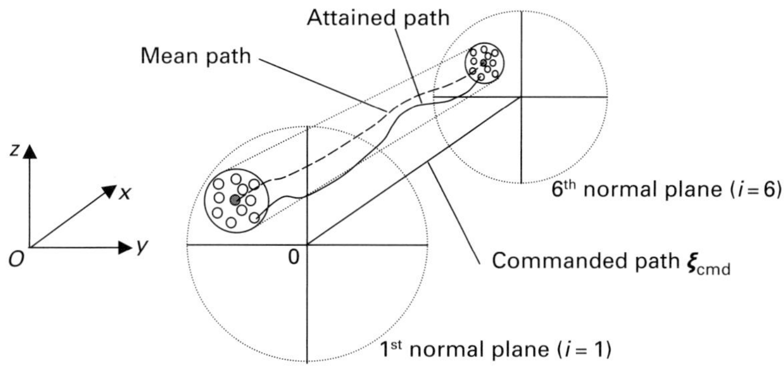

Figure 7.12 Path (positioning) accuracy and repeatability in three dimensions.

Each plane provides the y/z coordinate frame for computing one a_{P} and one r_{P} value. Several attained paths are measured, providing several measurement points per plane along ξ_{cmd}. These samples allow computation of the two-dimensional a_{P} and r_{P} for each plane using Equations (7.4) and (7.8). The x-position along the path is used as an additional third sample data point, resulting in the three-dimensional a_{T} and r_{T}. This is illustrated in Figure 7.12, where the point cloud is adopted from Figure 7.4. The measurement system's reference frame in Figure 7.12 is O-xyz. In practice the x-axis overlaps with ξ_{cmd}.

Path (positioning) accuracy

The path (positioning) accuracy a_{T} is defined as the maximum deviation between the commanded path and the mean path (the mean for N attained paths). Each attained path is measured at the same M positions along the commanded path. Use of different commanded paths is recommended in order that a representative portion of the workspace is covered. The j-th attained path has M sample points:

$$\xi_j(i) = \begin{bmatrix} x_j(i) \\ y_j(i) \\ z_j(i) \end{bmatrix}, \ j = 1, 2, \cdots, \ N, \ i = 1, 2, \cdots, \ M. \tag{7.21}$$

The mean path $\bar{\xi}$ is computed using the mean of three coordinates at each position along N attained paths:

$$\begin{gathered} \bar{\xi}(i) = \begin{bmatrix} \bar{x}(i) \\ \bar{y}(i) \\ \bar{z}(i) \end{bmatrix}, \ i = 1, 2, \cdots, \ M \\ \bar{x}(i) = \frac{1}{N}\sum_{j=1}^{N} x_j(i), \bar{y}(i) = \frac{1}{N}\sum_{j=1}^{N} y_j(i), \bar{z}(i) = \frac{1}{N}\sum_{j=1}^{N} z_j(i). \end{gathered} \tag{7.22}$$

The path (positioning) accuracy a_{T} is then defined in Equation (7.23) as the maximum mean error, that is, the Euclidean norm, between the commanded path and the mean path:

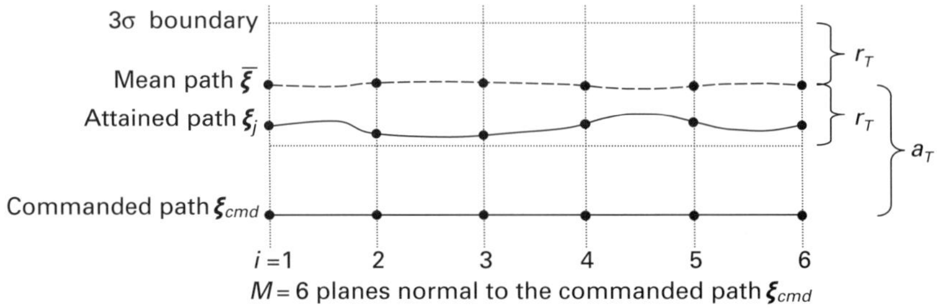

Figure 7.13 Definition of path accuracy and repeatability (2D view).

$$a_T = \max_i \|\bar{e}(i)\| \tag{7.23}$$

$$\begin{aligned}\bar{\boldsymbol{e}}(i) &= \bar{\boldsymbol{x}}(i) - \boldsymbol{x}_{cmd}(i) \\ &= \begin{bmatrix} \bar{x}(i) - x_{cmd}(i) \\ \bar{y}(i) - y_{cmd}(i) \\ \bar{z}(i) - z_{cmd}(i) \end{bmatrix} \\ &= \sqrt{(\bar{x}(i) - x_{cmd}(i))^2 + (\bar{y}(i) - y_{cmd}(i))^2 + (\bar{z}(i) - z_{cmd}(i))^2}.\end{aligned} \tag{7.24}$$

The attained positions used for computing the mean path are measured at M planes normal to the commanded path. Figure 7.13 illustrates the above definitions in two dimensions, showing the commanded path, the j-th attained path, and the mean path. Here $M = 6$ normal planes are selected for measurements.

The choice of M can affect the accuracy of a_T. It depends on the shape of the commanded path: a straight line will require fewer measurements than a 'curvy' path. Generally a curved path and 100% of the rated load and 50–100% of the rated velocity are recommended according to ISO standard 9283. However, a linear path and a reduced load (10% of the rated load) are also acceptable for testing path (positioning) accuracy.

Path (positioning) repeatability

The path (positioning) repeatability r_T characterizes a robot's ability to move its end-effector along a commanded path in the same direction N times. It is based on the positioning repeatability r_P in (7.18) and is defined as the maximum 3σ deviation from the mean path $\bar{\xi}(i)$, $i = 1, \cdots, M$ for all N attained paths $\xi_j(i)$, $j = 1, \cdots, N$, $i = 1, \cdots, M$. For the j-th path the distance (Euclidean norm) between the i-th measured position and the associated mean position is

$$d_j(i) = \sqrt{(x_j(i) - \bar{x}(i))^2 + (y_j(i) - \bar{y}(i))^2 + (z_j(i) - \bar{z}(i))^2}. \tag{7.25}$$

The mean distance at the i-th attained position is

$$\bar{d}(i) = \frac{1}{N}\sum_{j=1}^{N} d_j(i). \tag{7.26}$$

Example 7.6: accuracy and repeatability of a straight path

The path accuracy a_T and repeatability r_T of a cylindrical robot is tested in the Cartesian reference frame *O-xyz* of a measurement system. Consider the handling of a substrate along a straight end-effector trajectory. This applies to the handling of 300 mm wafers in and out of cassettes with tight tolerances: a 1.0 mm gap in FOUPs (SEMI standard E47.1). This requires good path accuracy near extended arm poses. The robot's *R*-axis is a commanded path ξ_{cmd} with six points, a straight line along the measurement system's *x*-axis with $y=0$ and $z=0$. See ξ_{cmd} in Figure 7.12 and Figure 7.13. The numerical example in Table 7.4 lists the commanded path ξ_{cmd}, a straight line along the *x*-axis, and the measured paths ξ_j, $j=1,..3$, all in *O-xyz* with $i=1,\ldots,6$. The ξ_j are measured at or near the commanded *x*-positions in *y/z* planes normal to the commanded path. The mean attained path $\bar{\xi}$ is computed using Equation (7.22). The mean distance $\bar{d}(i)$ and the standard deviation $\sigma(i)$ are computed using Equations (7.26) and (7.27), respectively. a_T and r_T are computed using Equations (7.23) and (7.29). □

Table 7.4. Path positioning accuracy and repeatability in Example 7.6.

	Path position i					
Parameter	$i=1$	$i=2$	$i=3$	$i=4$	$i=5$	$i=6$
$\xi_{cmd}(i)$	0.2 0 0	0.4 0 0	0.6 0 0	0.8 0 0	1 0 0	1.2 0 0
$\xi_{j=1}(i)$	0.193 −0.057 −0.015	0.371 −0.045 0.012	0.526 0.016 −0.101	0.826 0.017 0.038	0.965 0.034 −0.054	1.245 −0.106 0.014
$\xi_{j=2}(i)$	0.163 −0.039 0.008	0.383 0.049 −0.050	0.651 −0.024 0.003	0.820 0.056 0.031	0.986 −0.067 −0.034	1.217 −0.005 0.015
$\xi_{j=3}(i)$	0.255 0.012 0.008	0.420 0.061 0.072	0.549 0.010 0.029	0.787 0.125 0.043	0.958 0.041 0.035	1.238 −0.086 0.077
$\bar{\xi}(i)$	0.204 −0.028 0.000	0.392 0.022 0.027	0.575 0.001 −0.023	0.811 0.066 0.037	0.970 0.002 −0.018	1.233 −0.081 0.035
$\lVert\bar{e}(i)\rVert$	0.028	0.035	0.034	0.077	0.035	**0.094**
$\bar{d}(i)$	0.048	0.060	0.079	0.043	0.063	0.039
$\sigma(i)$	0.016	0.015	0.018	0.026	0.013	0.004
$\boldsymbol{a}_T$			**0.094**			
$\boldsymbol{r}_T$			**0.133**			

The standard deviation at the i-th position of the j-th attained path with respect to the mean path is

$$\sigma(i) = \sqrt{\frac{\sum_{j=1}^{N} \left(d_j(i) - \bar{d}(i)\right)^2}{N-1}}. \tag{7.27}$$

The path repeatability at the i-th position, given N attained paths, is

$$r_P(i) = \bar{d}(i) + 3\sigma(i). \tag{7.28}$$

The total path repeatability is the maximum of all individual positions,

$$r_T = \max_i \{r_P(i)\}. \tag{7.29}$$

7.4.2 Measurement system

An overview of a measurement system for testing a_T and r_T for SCARA-type, substrate-handling robots is illustrated in Figure 7.14. The measurement system comprises the usual hardware and software subsystems:

- Sensors: a 5.5 mm-wide 'curtain' of 1000 laser beams allows contact-free path measurements in two dimensions. The curtain is created by a precision-aligned laser transmitter/receiver pair. Table 7.5 lists the specifications for a suitable, commercially available laser sensor.
- Data acquisition system: hardware and software acquire the attained paths; the data acquisition system interfaces with the computer, which analyzes the data and stores the results.
- Data analysis: software transforms the data into useful information using the above equations.
- Test execution program: program automates the test by executing and coordinating all activities during the specified number of test cycles.

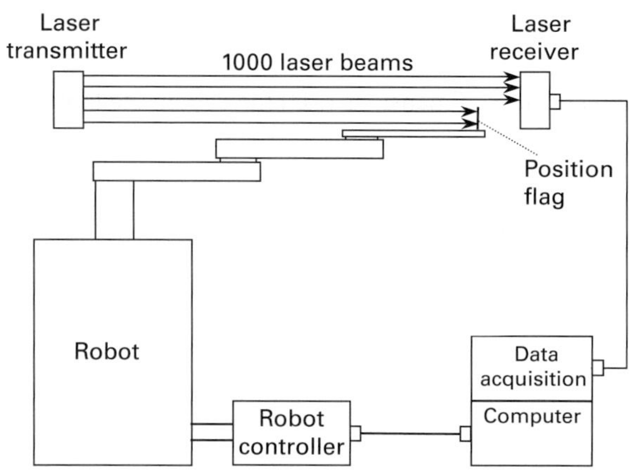

Figure 7.14 Measurement of vertical path (positioning) accuracy and repeatability.

Table 7.5. Technical data for a suitable laser sensor.

Feature	Specification
Light source	$800 \cdot 10^{-9}$ m (Pulse Class I Laser)
Resolution	$5.5 \cdot 10^{-6}$ m
Measurement distance (maximum)	0.35 m
Number of laser beams	1 000
Response time (maximum)	$1.4 \cdot 10^{-3}$ s
Data range (counts)	0 to 4 095
Data format	12 bit parallel

Example 7.7: path accuracy and repeatability of a wafer-handling robot

The above measurement system was used to test the arm of a 300 mm wafer-handling robot. Both a_T and r_T were determined. The kinematics and drive train of the SCARA-type, cylindrical robot are described in Example 5.3. The robot's *R*-axis for radial arm motion was aligned with the measurement system's *x*-axis. The commanded path of $M = 754$ points, $\xi_{cmd}(i)$ with $i = 1,\ldots, M$, is a straight line along the *x*-axis with $y = 0$ and $z = 0$. Measurements were taken in the horizontal *x*/*y* plane and in the vertical *x*/*z* plane (the laser curtain can be turned by 90° if needed). Figure 7.15a shows the acquired horizontal arm data, and Figure 7.15b shows the vertical data for the first of three attained paths, $\xi_1(i)$. The measurement resolution of 5.5 μm as listed in the sensor specifications (Table 7.5) can be seen in the figure. The *x*-data are encoder readings from the robot's *R*-axis. The figure shows a horizontal deviation from the commanded path of -72 μm $< y <$ 83 μm, and a vertical deviation of -72 μm $< z <$ 0 μm. The solid curve is the eight-point moving average of the sampled data.

The 'static' arm measurements in Figure 7.15 were taken at very slow motion to exclude dynamic effects, in particular arm vibration. The elapsed time for an arm extension was 10 seconds. The resulting a_T and r_T in Table 7.6 were computed using Equations (7.22) to (7.29). 3D data was obtained by combining the associated data in the horizontal *x*/*y* plane and the vertical *x*/*z* plane. In two dimensions (2D) the data describes the mechanical arm accuracy in the *y*/*z* plane, but excludes possible tracking errors along the *x*-axis. For three dimensions (3D) the data in the table includes such positions errors along the *x*-axis.

Figure 7.15 shows the a_T resulting from 2D data at $x = 0.249$ m, and from 3D data at $x = 0.098$ m. The full 3D error is not shown in the figure because the tracking error along the *x*-axis, which is largest near the start of motion, is not included.

The dynamic path performance at standard operating speeds is also of interest. Figure 7.16 compares the eight-point moving averages of the 'static' and dynamic attained paths. The arm oscillation is clearly visible and is likely due to several factors, including controller dynamics and the mass-spring dynamics of the arm links and belts. □

Table 7.6. Path accuracy and repeatability (μm) of the wafer-handling robot (Example 7.7).

	Performance in 2D (y,z)	Performance in 3D (x,y,z)
Path accuracy a_T	102	656
Path repeatability r_T	24	24

The high measurement resolution makes the test sensitive to vibration. The robot and measurement system are mounted on separate mechanical frames to prevent the transmission of robot vibration to the sensors. The measurement system's reproducibility can be tested using procedures that are defined, for example, in SEMI standard E89-1104.

The system configuration is illustrated in Figure 7.14 for vertical path positioning measurements. It can also be adjusted for horizontal displacement measurements by turning the laser curtain by 90°. The laser curtain between transmitter and receiver is interrupted at M measurement points by the position flag attached to the end-effector. The resulting 'shadow' is detected by the data acquisition system and interpreted as the flag's position. The test execution program running on the computer communicates with the robot controller and coordinates all robot motion and measurements.

7.5 Vibration analysis

This section presents a vibration analysis tool designed for the handling robots discussed in this book. Vibration refers to periodic or random mechanical oscillations about an equilibrium point (Thomson, 1996). It is almost always an undesirable phenomenon that wastes energy and creates unwanted effects including noise, accelerated wear, and inaccuracies in electromechanical equipment. Forced vibration occurs when an external force or motion is applied to a mechanical system. The external force or motion determines the vibration frequency, but the magnitude of the vibration is strongly dependent on the mechanical system itself. Free vibration occurs when a system vibrates freely at its natural frequencies until the vibration settles to zero. The level of vibration during motion is an important performance index for electromechanical systems such as robots. Vibration can reduce the product life of a robot and contribute to particle generation. In the case of robotic systems the vibration excitation is caused by a robot's drive train and the servo control loop. Excessive vibration can also limit the attainable speed and throughput performance of robots. It can cause substrate slippage on the end-effector ('wafer walk') and can affect system repeatability. Vibration must therefore be reduced to a level that does not negatively impact substrate handling and the process tool.

7.5.1 Algorithm

Robot arm vibration is observed as small structural displacements over time, superimposed on the commanded arm motion. Here the end-effector displacements in all three

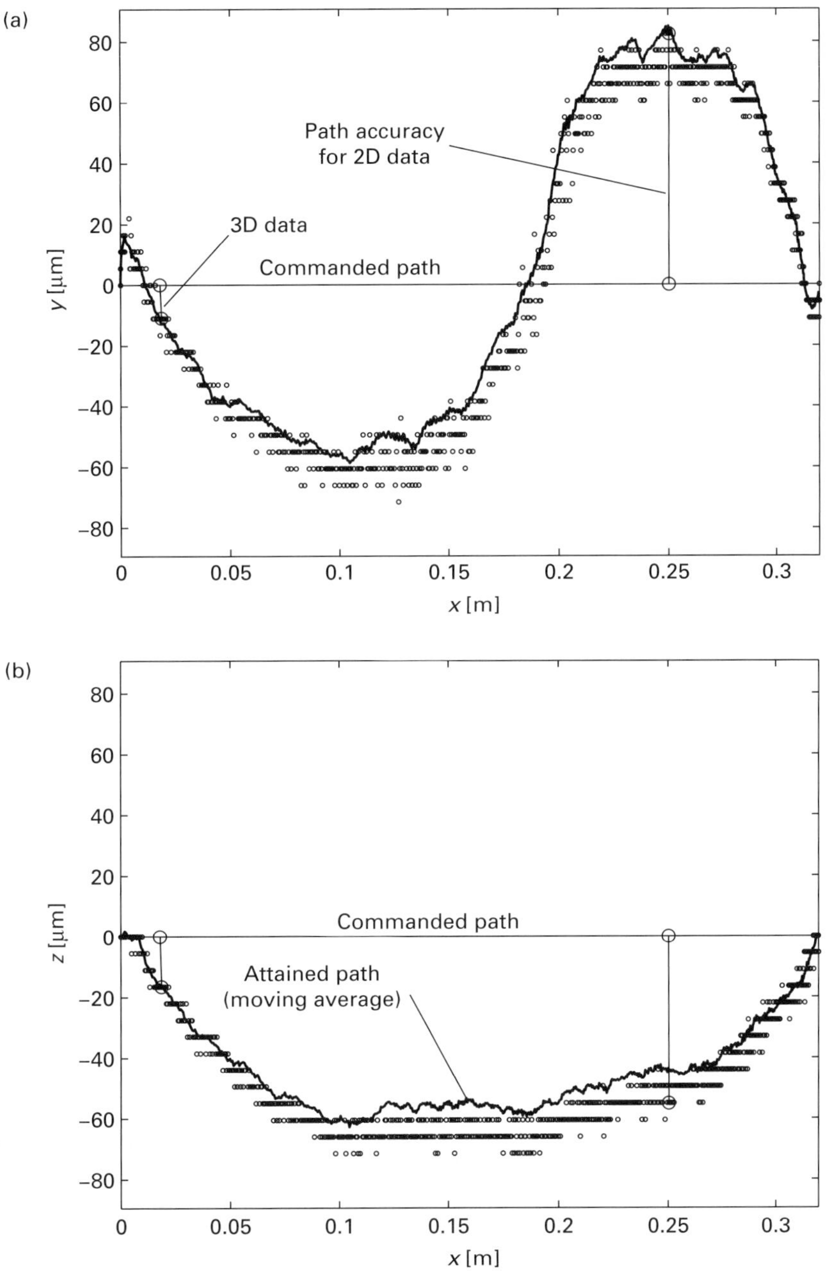

Figure 7.15 Path error in (a) the horizontal x/y plane, and (b) the vertical x/z plane. Note the measurement units (m, μm) (Example 7.7).

dimensions are of interest. Common performance criteria in vibration analysis are the peak value, mean value, mean square value, and root mean square (RMS) value of the infinite, continuous-time signal $\boldsymbol{x}(t)$. After data acquisition the finite, discrete-time signal

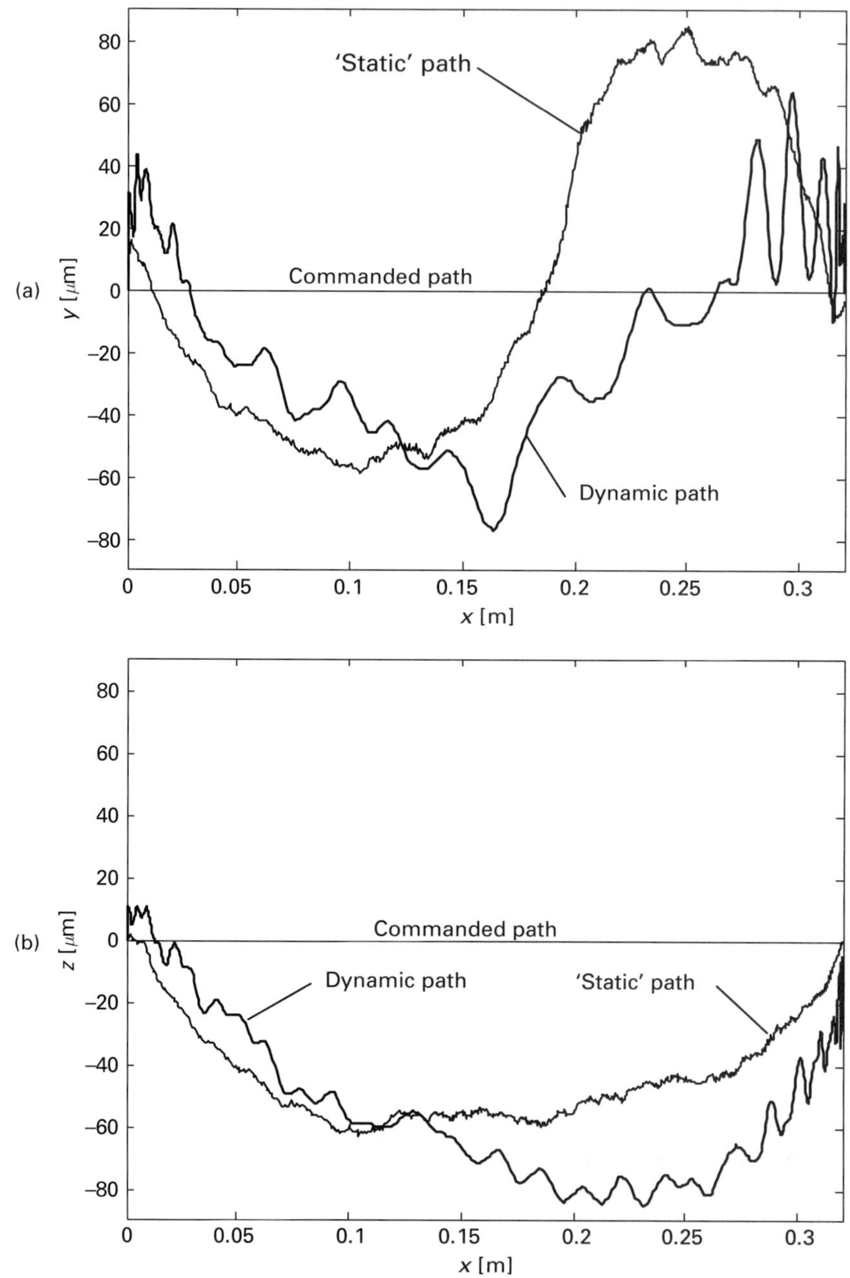

Figure 7.16 Static and dynamic performance in (a) the horizontal *x*/*y* plane, and (b) the vertical *x*/*z* plane. Note the measurement units (m, μm) (Example 7.7).

$\boldsymbol{x}(n)$, $n = 0, \ldots, N - 1$ is processed. $\boldsymbol{x}(n)$ denotes the signal at time $t = nT$, where T is the sampling period. The peak value is defined as the maximum absolute value of all $\boldsymbol{x}(n)$,

$$\hat{x} = \max(|x(n)|), \; n = 0, \ldots, \; N - 1 \tag{7.30}$$

The mean value is defined in Equation (7.3). The root mean square value is a measure of the energy in a vibration signal. Given N samples in the time series, the RMS value is defined as

$$x_{RMS} = \sqrt{\frac{1}{N}\sum_{n=0}^{N-1} x^2(n)}. \tag{7.31}$$

Vibration data analysis in the frequency domain is also used. The frequency domain representation of the finite-length time series $x(n)$ with N samples is the discrete Fourier transform (DFT) $X(k)$, which can be computed with the fast Fourier transform (FFT):

$$X(k) = \sum_{n=0}^{N-1} x(n) \cdot e^{-i2\pi k\frac{n}{N}}, k = 0, \ldots, \ N - 1. \tag{7.32}$$

Here i is the imaginary unit with $i^2 = -1$. Note that $x(n) = 0$ if n is not in the interval $n = 0,1,\ldots, N-1$, and that $X(k) = 0$ if k is not in the interval $k = 0,1,\ldots, N-1$ (Oppenheim and Schafer, 1989). Modern spectrum analyzers can compute the frequency response in real-time using the FFT. The frequency response of a system, usually represented by amplitude and phase, could also be obtained experimentally by measuring the system output using reference input signals of different frequencies but constant amplitude. A detailed discussion of vibration theory is presented, for example, in (Thomson, 1996). Here the magnitude $|X(k)|$ is used for robot vibration analysis, which permits an estimate of the robot arm's resonance frequencies, at which it oscillates at an increased amplitude.

Here the measured end-effector vibration data, that is, the sampled end-effector acceleration, is analyzed in both time and frequency domain. Acceleration is easy to measure and is the preferred signal for vibration analyses. In theory the relative displacement $x(t)$ can be obtained through double integration of the acceleration signal $\ddot{x}(t)$,

$$x(t) = \iint \ddot{x}(\tau)d\tau^2. \tag{7.33}$$

In practice, however, this method is susceptible to position errors and noise in $\ddot{x}(t)$. Errors from the double integration can accumulate quickly (Thong et al., 2004). In inertial navigation systems, for example, the position $x(t)$ is periodically corrected with an absolute position reference.

7.5.2 Vibration measurement system

Vibration is usually inferred from measured acceleration signals, for several reasons:

- Acceleration provides information directly related to the vibration velocity and amplitude.
- Sensors and instruments for measuring and analyzing the data are readily available at reasonable cost.
- The acceleration amplitude is of interest because it is the main source of undesired consequences of vibration, including damage and wear.

Accelerometers are simple micro-electromechanical systems (MEMS). Some consist of only a suspended cantilever beam and displacement-sensing circuitry. The total force acting on the sensor, including gravity, is measured. A typical vibration measurement system includes one or more accelerometers, a data acquisition system for sampling the signal and recording the sampled data, a signal analyzer, and a test execution program. The accelerometer produces an electrical signal proportional to the observed acceleration. Lightweight sensors are recommended to minimize their influence on the observed object, in this case the dynamics of the robot arm or end-effector under test. The signal analyzer computes the power spectrum, that is, the distribution of measured acceleration values across the frequency spectrum. Many commercial analyzers and software tools are available for this purpose.

The measurement system described here was developed as an integrated vibration test and analysis tool for the automated testing of substrate-handling robots. Two types of vibration are of interest:

- Forced vibration during robot motion from the initial to the commanded pose.
- Free vibration at the commanded end-effector position after the robot motion has been completed.

The block diagram in Figure 7.17 illustrates the hardware and software components of the PC-based vibration measurement system. The system utilizes three single-axis accelerometers, which are aligned with the measurement system's Cartesian coordinate frame *O-xyz*. The *x*, *y*, and *z* axes are aligned with the robot's *R*, Θ, and *Z* axes, respectively, and measure the vibration along these axes. The piezoelectric sensors have a sensitivity of 1 $V \cdot g^{-1}$, a range of ±5 g, and a bandwidth from 1 to 3000 Hz; they are powered by the sensor controller, and are attached to the end-effector. The controller

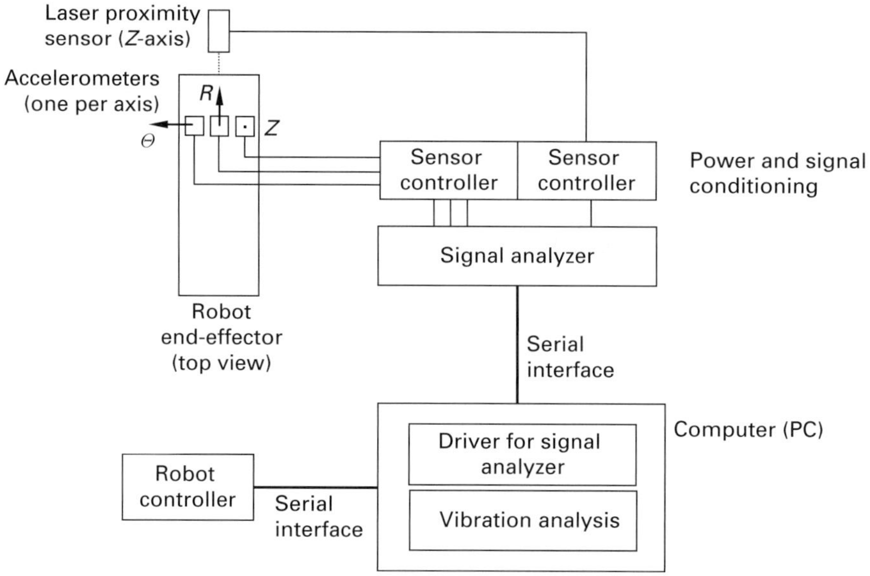

Figure 7.17 Block diagram of the vibration measurement system.

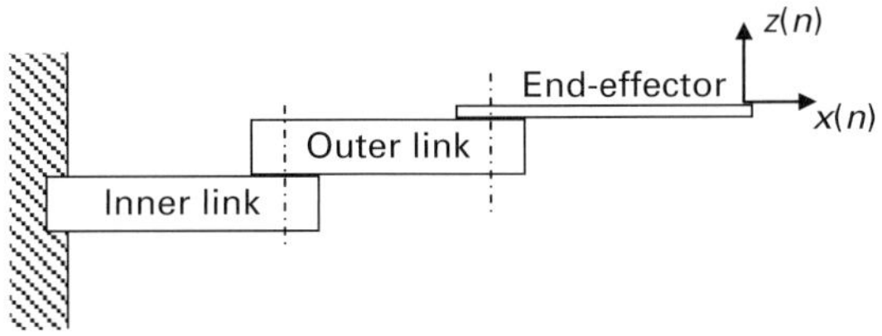

Figure 7.18 Rigidly mounted mechanical structure of a SCARA-type robot arm.

pre-processes the sensor signals and outputs a conditioned signal with a ±5 V output amplitude. The vertical, free end-effector vibration is measured with a laser proximity sensor (Figure 7.18). It measures the displacement from the end-effector's nominal Z-position and is located above the commanded end-effector position. For that purpose the arm is considered firmly mounted at the shoulder joint. The proximity sensor has a reference distance of 40 mm, a measurement range of ±5 mm, and a resolution of 2 μm. The signal is conditioned and pre-processed by a separate sensor controller that has a bandwidth of 900 Hz. All four sensor signals are further processed by a signal analyzer. The signal analyzer has four input channels and acquires, processes, and analyzes all accelerometer and position signals. It is configured with a unity gain and a low pass cut-off frequency of 10 kHz. The analyzer performs the spectrum analysis and also provides useful 'virtual instruments,' software tools that mimic oscilloscope, spectrum analyzer, and network analyzer functions. The test execution program on the PC coordinates and synchronizes the data acquisition with the commanded robot motion. All data is eventually transferred to the PC and saved for further use. The test execution program coordinates the tasks of the various system components and automates the test process and data analysis.

7.5.3 Data analysis and performance indices

The sampled data is analyzed in time and frequency domains. The DFT is scaled using the number of sample points, N:

$$A(k) = \frac{|X(k)|}{N} = \frac{\sqrt{(\mathrm{Re}(X(k)))^2 + (\mathrm{Im}(X(k)))^2}}{N}. \tag{7.34}$$

Here $N = 4096$. The time series and the DFT provide the data to establish common vibration performance indices for robots, including the following:
Time domain:

- The maximum acceleration (absolute value) during the commanded motion
- The acceleration RMS during the commanded motion, an indicator of the energy level in the vibration signal

Example 7.8: vibration analysis of a wafer-handling robot

This example presents the vibration test results for an atmospheric wafer-handling robot. Figure 7.19 shows the end-effector acceleration for the Θ and R axes of a 300 mm wafer-handling robot during forced and free vibration. The graphs (a) and (b) show the time series while the graphs (c) and (d) show the scaled DFT per Equation (7.34). The commanded motion is a straight line along the radial R-axis in the robot's cylindrical coordinate frame, from the initial pose near the Z-axis ($R=0$) to the commanded pose near the outer workspace limit. The commanded, trapezoidal acceleration profile is shown in Figure 7.19b for comparison with the actual, measured profile. Note the correlation between the commanded acceleration profile and the oscillation along the R-axis.

The timing for the measurements was:

$t_0 = 0.00$ s: begin data acquisition
$t_1 = 0.24$ s: begin robot motion
$t_2 = 0.82$ s: end robot motion
$t_3 = 2.20$ s: end data acquisition.

The sampling frequency was set to 1280 Hz, 2.56 times the bandwidth of the frequency analyzer (500 Hz). Note that the dominant frequencies are at or below 20 Hz, which is a typical range for the tested robot type and similar electromechanical systems. □

- Maximum vertical end-effector displacement during settling at the final position
- The stabilization time needed for the vibration to settle below a specified displacement or acceleration tolerance (ISO standard 9283).

Frequency domain:

- The RMS per frequency band. Here the bands were selected as 0 – 8 Hz (low frequency band), 9 – 100 Hz (medium frequency band), and above 100 Hz (high frequency band)
- The first and second dominant frequencies with amplitude peak values in the combined low and medium bands (0 – 100 Hz)
- The 'Q value' of a dominant frequency, which indicates how narrow the peak of that frequency is. Q is the unitless ratio of the resonance frequency to the –3 dB bandwidth of the resonance peak.

The performance indices to be used depend on specific application requirements.

7.6 Mechanical axis decoupling

7.6.1 Motivation

The benefits of mechanically decoupled axes of motion have been discussed in Section 5.4.4 in the context of cylindrical robots for substrate handling. The end-effector

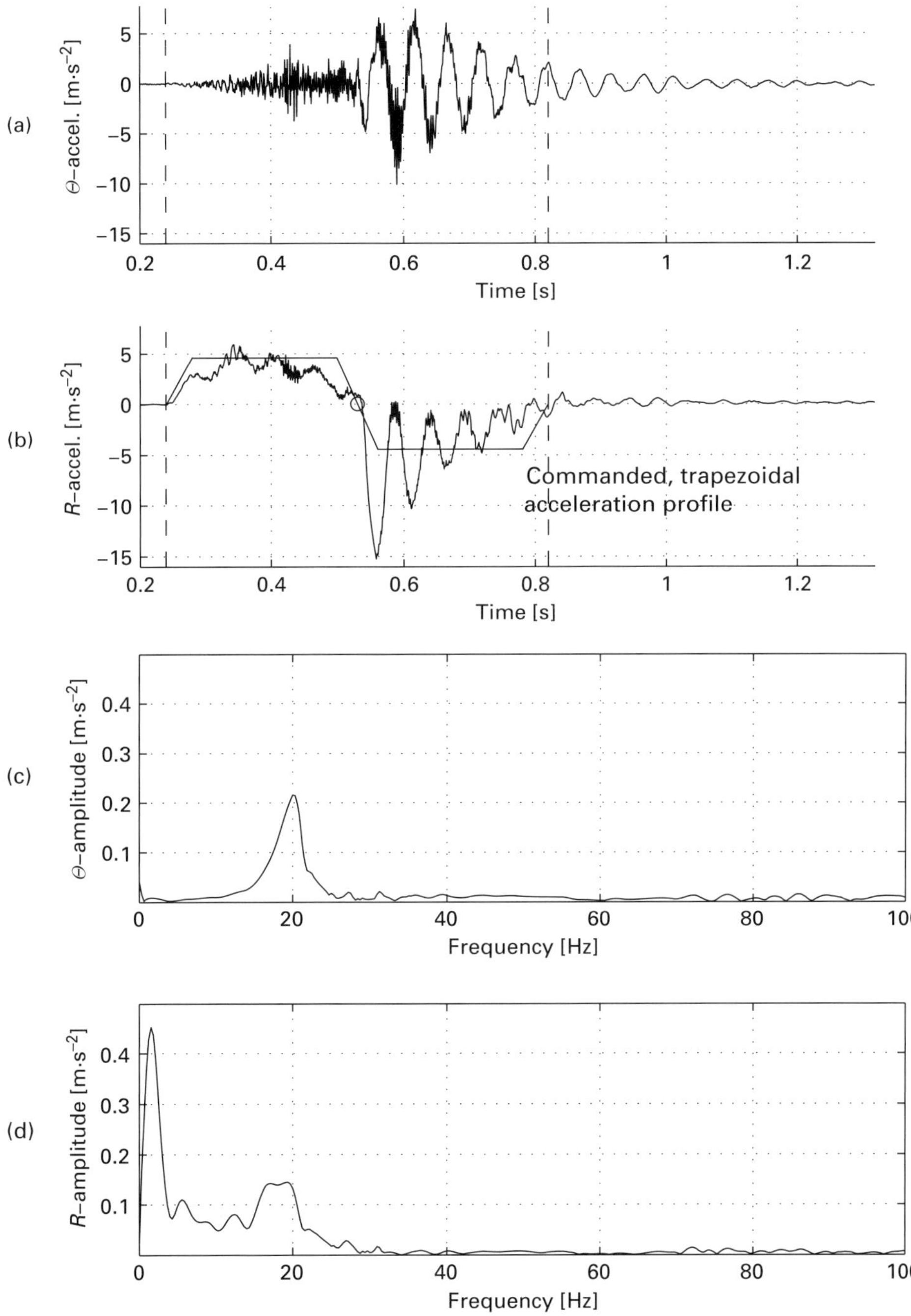

Figure 7.19 Measured arm vibration in time and frequency domains.

trajectory for vertical Z-motion is, in theory, perpendicular to the horizontal robot mounting plane. In practice, however, it deviates from the ideal path during vertical motion, which is acceptable within a specified tolerance. For example, the tight tolerances in 300 mm wafer cassettes require that the extended robot arm does not drift more than about 0.5 mm in Θ during vertical Z-motion. In this case the Θ and Z axes are considered decoupled. SEMI standard E1.9 specifies casette designs for 13 to 25 300 mm wafers, with a 10 mm pitch between adjacent wafer slots. Figure 7.20 is an oblique view of the maximum dimensions of a cassette for 13 wafers. For such a cassette the reference position used for 'teaching' a robot position is usually the wafer center position in the bottom slot. The R, Θ, and Z coordinates of that position are stored and used for loading and unloading substrates. The typical vertical dimension between the bottom and top wafer is 250 mm, and can reach 400 mm for some applications. A drift in Θ during Z-motion beyond the specified tolerance can result in contact between the handled wafer and the cassette wall.

Example 7.9: 300 mm wafer carrier

This example presents tolerances in a standardized 300 mm wafer carrier. The front-opening unified pod (FOUP), a non-removable wafer cassette in an enclosure, is an example of a wafer carrier used in the front-end semiconductor industry. Its front-opening mechanical interface mates with an automated door opener as specified by SEMI standards E47.1 and E1.9. The outer edge of the wafer pick-up volume, the cassette wall, is of particular interest: the resulting wafer pick-up volume in a FOUP measures 151 mm from the nominal wafer center line for all 25 wafers. This leaves a 1.0 mm tolerance between a 300 mm wafer and the FOUP wall for robot positioning errors. Axis decoupling is achieved if the following two conditions are met:

- The Z-axis is perpendicular to the robot mounting plane, the reference plane for both the robot and the FOUP.
- The path positioning accuracy for the Z-axis is better than 0.5 mm for the vertical cassette dimension, leaving a minimum safety margin of 0.5 mm.

Substrate-handling applications involve at least one substrate carrier and one tool grouped around the robot. Decoupled axes at the specified path accuracy are required for the applicable portion of the robot's workspace. Several tests that represent that portion of the workspace must be performed. Testing at only a single location would evaluate the path accuracy only for that commanded path. □

7.6.2 Measurement system

The measurement system described here was designed for testing the decoupling of the Θ and Z axes of cylindrical robots. The main steps of a test are to establish a three-dimensional reference frame, to align the robot's coordinate frame with the reference frame, and to take the measurements.

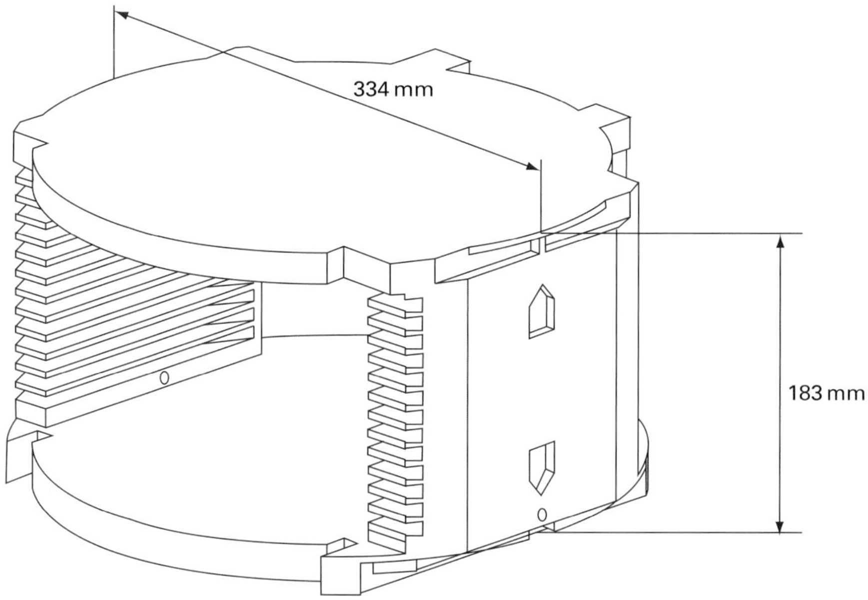

Figure 7.20 Oblique view of the maximum dimensions of a 300 mm wafer cassette (SEMI standard E1.9).

Overview

The laser-based measurement system includes the following hardware and software:

- Sensors: two contact-free laser position sensors with a ±1 μm position resolution and maximum measurement distance of 0.5 m. The lasers produce a beam with a 0.8 mm diameter. The detectors have an 8 mm diameter and a ±25 μm position accuracy over the sensor area.
- Data acquisition system: a PC with several software tools connects to the sensors via a USB bus and acquires the detected laser-beam position.
- Data analysis software: the software runs on the PC, transforming the acquired data into metric position data and displaying it graphically.
- Test execution software: the software automates the test and executes the required number of tests.

Figure 7.21 shows the configuration of the measurement system with a SCARA-type handling robot. The objective is to measure:

- Whether the robot's vertical *Z*-axis is perpendicular to the mounting plane
- Whether the vertical *Z*-motion of the extended arm is parallel to the robot's *Z*-axis.

Ideally any vertical trajectory is parallel to the robot's *Z*-axis for the entire workspace. Therefore the two laser beams in Figure 7.21 must be parallel to one another and perpendicular to the horizontal robot mounting plane. Vertical *Z*-motion is measured at a specified R/Θ-position in the cylindrical coordinate frame.

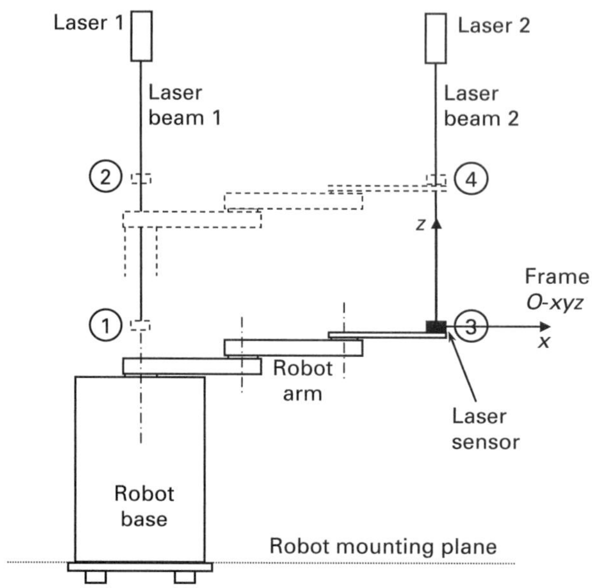

Figure 7.21 Measurement system configuration for the axis decoupling test.

First, the measurement system is calibrated such that both laser beams are vertical and parallel. Second, the robot *Z*-axis is aligned with laser beam 1. Third, laser beam 2 is aligned parallel to laser beam 1. Laser beam 2 represents the commanded path at an extended arm position and is thus used to determine the path positioning accuracy. The procedure is described below.

Calibrating the measurement system

Calibrating the measurement system involves three steps. The procedure does not involve the robot. The reference frame for calibrating the measurement system is Earth's gravity: the vertical *z*-axis of the measurement system's reference frame *O-xyz* is aligned with the gravitational field:

- Step 1. Level the robot mounting surface.
- Step 2. Calibrate laser beam 1 (LB_1) perpendicular to the robot mounting plane.
- Step 3. Calibrate laser beam 2 (LB_2) parallel to LB_1.

Step 1, leveling of the robot mounting surface, can be accomplished with conventional leveling tools, for example a bubble level. Granite surface plates are good (but expensive) mounting surfaces for measurement systems. They ensure precise flatness, and their heavy weight minimizes the risk of displacing an already leveled surface.

Step 2, calibrating the laser beam LB_1 vertically, parallel to the gravitational field and therefore perpendicular to the horizontal mounting plane, is accomplished by reflecting LB_1 using a level mirror or a reflective fluid on the ground. LB_1 is vertical with respect to gravity if the reflected beam returns to its source, laser 1.

Step 3, calibrating laser beam LB_2 parallel to LB_1, is accomplished by repeating the procedure in Step 2 for laser 2. LB_1 and LB_2 are now parallel and vertical.

The reproducibility of the measurement system, that is, the repeatability of test data under nominal conditions, can be tested using a procedure like the one defined in SEMI standard E89-1104.

7.6.3 Measurement and data analysis

After installing the robot in the calibrated measurement system its path positioning accuracy is measured with respect to the laser beams LB_1 and LB_2: the beams are used as two commanded vertical paths, while the laser sensor is attached to the end-effector. The circled numbers 1 to 4 in Figure 7.21 indicate the initial and final end-effector positions on the vertical paths along LB_1 (from point P_1 to P_2) and LB_2 (from point P_3 to P_4). The end-effector positions are acquired in the measurement system's reference frame *O-xyz* (Figure 7.22c). In the robot's cylindrical coordinate frame, the four end-effector positions are:

$$\begin{aligned} &\mathrm{P}_1 : \Theta_1 = \Theta_{user}, R_1 = 0, Z_1 = 0 \\ &\mathrm{P}_2 : \Theta_2 = \Theta_{user}, R_2 = 0, Z_2 = Z_{\max} \\ &\mathrm{P}_3 : \Theta_3 = \Theta_{user}, R_3 = R_{\max}, Z_3 = 0 \\ &\mathrm{P}_4 : \Theta_4 = \Theta_{user}, R_4 = R_{\max}, Z_4 = Z_{\max}. \end{aligned} \tag{7.35}$$

The laser sensor detects the horizontal laser position in the *x*/*y* coordinate frame and combines it with the robot *Z*-position to obtain a 3D path. Θ_{user} is a Θ-position specified by the user. The path accuracy a_T, defined in ISO standard 9283 and discussed in Section 7.4, is a suitable performance measure for evaluating the axis decoupling, i.e., tracking the commanded end-effector path LB_2. ISO standard 9283 recommends testing the path accuracy for a representative portion of the workspace. This is accomplished here with a series of decoupling tests at several Θ-positions, for example $\Theta_{\text{user}} = 0^\circ, 60^\circ, \ldots, 300^\circ$. It is economical to use the above measurement system with two laser sensors and to turn the entire robot to all Θ_{user} for a series of measurements, rather than building a complex measurement system with laser beams installed at all Θ_{user}.

Figure 7.22 shows a top view of the robot's kinematic skeleton structure. The laser sensor is attached to the end-effector. The retracted arm in Figure 7.22a corresponds to the commanded path from P_1 to P_2. The extended arm in Figure 7.22b corresponds to the commanded path from P_3 to P_4. Figure 7.22c illustrates the laser sensor's horizontal *x*/*y* coordinate frame. The calibrated zero reference point is the origin of the crosshairs. The *i*-th laser measurement is shown as point (x_{Pi}, y_{Pi}). The *x*-axis is aligned with the robot's *R*-axis.

The measurement procedure involves these main steps:

- Step 1. Align the robot *Z*-axis with laser beam LB_1 from P_1 to P_2. This ensures a vertical *Z*-axis and perpendicularity to the horizontal mounting plane.
- Step 2. Determine the robot's axis decoupling along LB_2 from P_3 to P_4 for $\Theta_{\text{user}} = 0^\circ$, $60^\circ, \ldots, 300^\circ$. Other Θ_{user} values are also suitable.

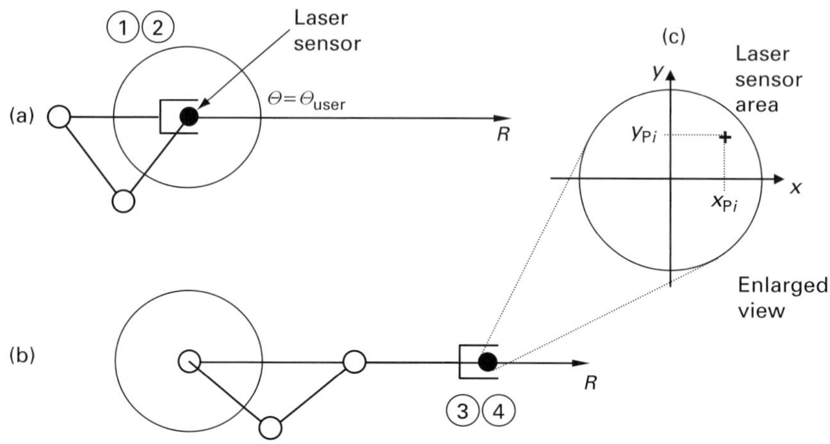

Figure 7.22 Axis decoupling measurement set-up (top view).

- Step 3. Compute the path positioning accuracy for different values of Θ_{user} using Equation (7.23).

Step 1, aligning the robot Z-axis with laser beam LB_1, is achieved by repeatedly measuring the end-effector positions along LB_1. The robot is aligned if the computed path positioning accuracy remains within the specified tolerance. Otherwise the robot must be reworked and this test repeated.

Step 2, determining the robot's axis decoupling, is achieved by commanding a slow, vertical motion along LB_2 and acquiring the end-effector positions at a suitable rate.

Step 3, computing the path positioning accuracy for different Θ_{user} values, is achieved using the ISO 9283 algorithms in Section 7.4. If the axis decoupling does not meet the robot's specifications it must be reworked and Steps 2 and 3 repeated.

7.7 Applicable and related standards

Several industry standards and guidelines apply directly or indirectly to the robots discussed in this chapter. The following list provides a selection. ANSI standards are published by the American National Standards Institute (ANSI). IEST standards are published by the Institute of Environmental Sciences and Technology (IEST). ISO standards are published by the International Organization for Standardization (ISO). RIA standards are published by the Robotic Industries Association (RIA). Sematech standards are published by International Sematech. SEMI standards are published by Semiconductor Equipment and Materials International (SEMI). Contact information for these organizations is listed in Appendix B.

ANSI/RIA R15.05-2, Industrial Robots and Robot Systems – Path-Related and Dynamic Performance Characteristics – Evaluation.

IEST-RP-CC001.4, HEPA and ULPA filters.

ISO 9283:1998(E), Manipulating industrial robots – Performance criteria and related test methods.

ISO 9787:1999, Manipulating industrial robots – Coordinate systems and motion nomenclatures.

ISO 9946:1999, Manipulating industrial robots – Presentation of characteristics.

ISO 10218–1:2006, Robots for industrial environments – Safety requirements – Part 1: Robot.

ISO vocabulary 1993, International Vocabulary of Basic and General Terms in Metrology.

Sematech ITRS 2003, The International Technology Roadmap for Semiconductors.

Sematech 92051107A-STD, Sematech guide for contamination control in design, assembly, and delivery of semiconductor manufacturing equipment, technology transfer.

SEMI E1.9, Mechanical specification for cassettes used to transport and store 300 mm wafers.

SEMI E10, Specification for Definition and Measurement of Equipment Reliability, Availability, and Maintainability (RAM).

SEMI E21-94, Cluster Tool Module Interface: Mechanical Interface and Wafer Transport Standard.

SEMI E89-1104, Guide for Measurement System Analysis.

SEMI F21-1102, Classification of airborne molecular contaminant levels in clean environments.

SEMI M52-0704, Guide for specifying scanning surface inspection systems for silicon wafers for the 130 nm technology generation.

Appendix A: SI units and conversion tables

A.1 SI units

International System of Units (SI Units) is the modern form of the metric system and is the official system of measurement in almost all countries worldwide (with the exception of the USA, Liberia, and Myanmar). The SI base units are listed in Table A.1. Some derived SI units, expressed in SI base units with no numerical factor other than 1, are listed in Table A.2.

Commonly used units also include:

Length: 1 centimeter (cm) = 10^{-2} m, 1 millimeter (mm) = 10^{-3} m, 1 micrometer (μm) = 10^{-6} m, 1 nanometer (nm) = 10^{-9} m

Angle: radian (rad), degree (°)

Angular velocity: degrees per second ($°\cdot s^{-1}$), radians per second ($rad\cdot s^{-1}$)

A.2 Unit conversion tables

The following examples illustrate the use of the unit conversion tables.

Pressure units

SI-based unit: Pa

Example: 1 Pa = 10^{-5} bar = $9.869 \cdot 10^{-6}$ atm = $7.501 \cdot 10^{-3}$ torr = $1.45 \cdot 10^{-4}$ psi

Leak rate units

SI-based unit: $Pa\cdot m^3\cdot s^{-1}$

Example: 1 $Pa\cdot m^3\cdot s^{-1}$ = 10 $mbar\cdot liter\cdot s^{-1}$ = 9.87 $atm\cdot cm^3\cdot s^{-1}$ = 7.501 $torr\cdot liter\cdot s^{-1}$

Pumping speed units

SI-based unit: $m^3\cdot s^{-1}$

Example: 1 $m^3\cdot s^{-1}$ = 3600 $m^3\cdot h^{-1}$ = 1000 $liter\cdot s^{-1}$

Outgassing rate units

SI-based unit: $Pa\cdot m^3\cdot m^{-2}\cdot s^{-1} = Pa\cdot m\cdot s^{-1}$

Example: 1 $Pa\cdot m\cdot s^{-1}$ = 0.001 $Pa\cdot liter\ m^{-2}\cdot s^{-1}$

Table A.1. The seven SI base units.

Quantity	Name	Symbol
Length	meter	m
Mass	kilogram	kg
Time	second	s
Electric current	ampere	A
Absolute temperature	kelvin	K
Amount of substance	mole	mol
Luminous intensity	candela	cd

Table A.2. Derived SI base units.

Quantity	Name	Symbol, SI base units
Velocity	–	$1\ m\cdot s^{-1}$
Force, weight	newton	$1\ N = 1\ m\cdot kg\cdot s^{-2}$
Energy, work	joule	$1\ J = 1\ m^2\cdot kg\cdot s^{-2}$
Power	watt	$1\ W = 1\ m^2\cdot kg\cdot s^{-3}$
Voltage	volt	$1\ V = 1\ m^2\cdot kg\cdot s^{-3}\cdot A^{-1}$
Electric resistance	ohm	$1\ \Omega = 1\ m^2\cdot kg\cdot s^{-3}\cdot A^{-2}$
Pressure	pascal	$1\ Pa = 1\ m^{-1}\cdot kg\cdot s^{-2}$

Table A.3. Pressure units and conversion factors.

	Pa	bar	atm	torr	psi
Pa	1	$1.0 \cdot 10^{-5}$	$9.869 \cdot 10^{-6}$	$7.501 \cdot 10^{-3}$	$1.45 \cdot 10^{-4}$
bar	$1.0 \cdot 10^{5}$	1	$9.869 \cdot 10^{-1}$	$7.501 \cdot 10^{2}$	$1.450 \cdot 10^{1}$
atm	$1.013 \cdot 10^{5}$	1.013	1	760	$1.470 \cdot 10^{1}$
torr	$1.333 \cdot 10^{2}$	$1.333 \cdot 10^{-3}$	$1.316 \cdot 10^{-3}$	1	$1.934 \cdot 10^{-2}$
psi	$6.895 \cdot 10^{3}$	$6.895 \cdot 10^{-2}$	$6.805 \cdot 10^{-2}$	$5.171 \cdot 10^{1}$	1

Table A.4. Leak rate units and conversion factors.

	$Pa\cdot m^3\cdot s^{-1}$	$mbar\cdot liter\cdot s^{-1}$	$atm\cdot cm^3\cdot s^{-1}$	$torr\cdot liter\cdot s^{-1}$
$\mathbf{Pa\cdot m^3\cdot s^{-1}}$	1	10	9.87	7.501
$\mathbf{mbar\cdot liter\cdot s^{-1}}$	0.1	1	0.987	0.750
$\mathbf{atm\cdot cm^3\cdot s^{-1}}$	0.101	1.01	1	0.758
$\mathbf{torr\cdot liter\cdot s^{-1}}$	0.133	1.333	1.319	1

Table A.5. Pumping speed units and conversion factors.

	$m^3 \cdot s^{-1}$	$m^3 \cdot h^{-1}$	$liter \cdot s^{-1}$	$liter \cdot min^{-1}$	$ft^3 \cdot min^{-1}$
$\mathbf{m^3 \cdot s^{-1}}$	1	3600	1000	$6 \cdot 10^4$	2119
$\mathbf{m^3 \cdot h^{-1}}$	$2.778 \cdot 10^{-4}$	1	0.278	16.667	0.588
$\mathbf{liter \cdot s^{-1}}$	0.001	3.6	1	60	2.118
$\mathbf{liter \cdot min^{-1}}$	$1.667 \cdot 10^{-5}$	0.06	$1.667 \cdot 10^{-2}$	1	$3.53 \cdot 10^{-2}$
$\mathbf{ft^3 \cdot min^{-1}}$	$4.717 \cdot 10^{-4}$	1.7	0.472	28.31	1

Table A.6. Outgassing rate units and conversion factors.

	$Pa \cdot m \cdot s^{-1}$	$Pa \cdot liter \cdot m^{-2} \cdot s^{-1}$	$torr \cdot liter \cdot cm^{-2} \cdot s^{-1}$	$W \cdot m^{-2}$
$\mathbf{Pa \cdot m \cdot s^{-1}}$	1	0.001	$7.501 \cdot 10^{-4}$	1
$\mathbf{Pa \cdot liter \cdot m^{-2} \cdot s^{-1}}$	1000	1	$7.501 \cdot 10^{-7}$	1000
$\mathbf{torr \cdot liter \cdot cm^{-2} \cdot s^{-1}}$	1333	$1.333 \cdot 10^6$	1	1333
$\mathbf{W \cdot m^{-2}}$	1	0.001	$7.501 \cdot 10^{-4}$	1

Appendix B: Standards organizations

ASTM International
100 Barr Harbor Drive West Conshohocken, PA 19428, USA Tel: +1-610-832-9500 Website: http://www.astm.org

Electrostatic Discharge Association (ESDA)
7900 Turin Road, Building 3 Rome, NY 13440-2069, USA Tel: +1-315-339-6937 Website: http://www.esda.org

International Organization for Standardization (ISO)
1, ch. de la Voie-Creuse, Case postale 56 CH-1211 Geneva 20, Switzerland Tel: +41-227490111 Website: http://www.iso.org

International Federation of Robotics (IFR)
c/o VDMA Robotics + Automation Lyoner Str. 18 D-60528 Frankfurt, Germany Tel: +49-69-66031502 Website: http://www.vdma.org

Robotic Industries Association (RIA)
P.O. Box 3724 Ann Arbor, MI 48106, USA Tel: +1-734-994-6088 Website: http://www.robotics.org

International Sematech
2706 Montopolis Drive Austin, TX 78741, USA Tel: +1-512-356-3500 Website: http://www.sematech.org

CAN in Automation (CiA)
Kontumazgarten 3 D-90429 Nürnberg, Germany Tel: +49-911-9288190 Website: http://www.can-cia.org

American National Standards Institute (ANSI)
1819 L Street, NW, 6th floor Washington, DC 20036, USA Tel: +1-202-293-8020 Website: http://www.ansi.org

International Electrotechnical
Commission (IEC) Rue de Varembé 3 CH-1211 Geneva 20, Switzerland Tel: +41-22-9190211 Website: http://www.iec.ch

Institute of Environmental Sciences and Technology (IEST)
2340 S. Arlington Heights Road, Suite 100 Arlington Heights, IL 60005, USA Tel: +1-847-981-0100 Website: http://www.iest.org

International Union of Vacuum Science Technique and Applications
Avenue de la Renaissance 30 B-1000 Brussels, Belgium Website: http://www.iuvsta.org/

Japan Robot Association (JARA)
Kikaishinko Building 3-5-8 Shibakoen, Minato-ku Tokyo 105-0011, Japan Tel: +81-3-34342919 Website: http://www.jara.jp

Semiconductor Materials and Equipment Industry (SEMI)
3081 Zanker Road San Jose, CA 95134, USA Tel: +1-408-943-6900 Website: http://www.semi.org

SAE International (Society of Automotive Engineers)
400 Commonwealth Drive Warrendale, PA 15096, USA Tel: +1-724-776-4841 Website: http://www.sae.org

Appendix C: Standard temperature and pressure (STP)

Standard temperature and pressure (STP) is a standard set of conditions for experimental measurements, enabling comparisons to be made between sets of data. Note that 0 °C is equivalent to 273.15 K.

Table C.1. Common standard conditions for experimental measurements.

Temperature (°C)	Absolute pressure (kPa)	Relative humidity (%)	Standards organization
0	100.000		International Union of Pure and Applied Chemistry (IUPAC)
0	101.325		National Institute of Standards and Technology (NIST), International Organization for Standardization (ISO 10780)
15	101.325	0	ICAO's International Standard Atmosphere (ISA), ISO 13443, European Environment Agency (EEA), Electricity and Gas Inspection Act of Canada (EGIA)
20	101.325		United States Environmental Protection Agency (EPA), National Institute of Standards and Technology (NIST)
25	101.325		United States Environmental Protection Agency (EPA)
25	100.000		Standard Ambient Temperature and Pressure (SATP)
20	100.000	0	Compressed Air and Gas Institute (CAGI)
15	100.000		Society of Petroleum Engineers (SPE)

References

Aalund, M. P. (1998). Controlling wafer transport in a vacuum environment. *MICRO Magazine*, June 1998, 93–104. Retrieved from the Internet on June 5, 2006 at http://www.micromagazine.com/archive/98/06/aalund.html.

Aalund, M. P. and Mathia, K. (2001). Emerging trends in fab automation. *Silicon Futures*, December 2001, 30–33.

Aalund, M. P. and Mathia, K. (2007). Surface finishes for capital equipment in industrial cleanroom environments. *Semiconductor International*, December 2007. Retrieved from the Internet on December 3, 2007 at http://www.semiconductor.net.

Accuratus (2009). Aluminum oxide, Al_2O_3. Retrieved from the Internet on January 19, 2009 at http://www.accuratus.com/alumox.html.

Alentiev, A. and Yampolskii, Y. (2006). Prediction of gas permeation parameters of polymers. In *Materials Science of Membranes for Gas and Vapor Separation*, ed. B. Freeman, Y. Yampolskii, and I. Pinnau. New York: Wiley InterScience.

Alwitt, R. S. (2002). Anodizing. *Electrochemistry Encyclopedia*, Cleveland, OH: Case Western Reserve University. Retrieved from the Internet on February 19, 2006 at http://electrochem.cwru.edu/encycl/art-a02-anodizing.htm.

Aron, P. (1981). Robots revisited – One year later. In Congress of the United States Office of Technology Assessment, *Exploratory Workshop on the Social Impacts of Robotics: Summary and Issues*, February 1982. Retrieved from the Internet on February 28, 2008 at http://fas.org/ota/otareports/year/1982a.

Asada, H. and Slotine, J. J. E. (1986). *Robot Analysis and Control*. New York: Wiley InterScience.

Ascher, U. M., Pai, D. K., and Cloutier, B. P. (1997). Forward dynamics, elimination methods, and formulation stiffness in robot simulation. *International Journal of Robotics Research*, **16**, 749–758.

Asimov, I. (1950). *I, Robot*. New York: Gnome Press.

Åström, K. and Hägglund, T. (1994). *PID Controllers: Theory, Design, and Tuning*, 2nd edn. Research Triangle Park, NC: Instrument Society of America.

Belt Technologies Inc. (1999). *Design Guide and Engineer's Reference for Metal Belts*. Product brochure. Durham, UK: Belt Technologies, Inc.

Björklöf, D. (1999). Immunity testing: Examining requirements and test methods. *Compliance Engineering*. Annual Reference Guide, European Edition, 1999, 51. Retrieved from the Internet on March 28, 2008 at http://www.ce-mag.com.

Campbell, W. A. and Scialdone, J. J. (1997). *Outgassing Data for Selecting Spacecraft Materials*. Reference publication 1124, Rev. 4. Linthicum Heights, MD: NASA Center for AeroSpace Information.

Carlson, A., Le, T., and Hallen, J. (2007). Improving wafer yields with integrated all-surface inspection. *Solid State Technology*, June 2007.

Chambers, A., Fitch, R. K., and Halliday, B. S. (1998). *Basic Vacuum Technology*, 2nd edn. London: Taylor & Francis.

Chase, G. (2008). Ohms per square what! *ESD Journal* (online). Retrieved from the Internet on March 28, 2008 at http://www.esdjournal.com/techpapr/ohms.htm.

Christensen, C. M., King, S., Verlinden, M., and Yang, W. (2008). The new economics of semiconductor manufacturing. *IEEE Spectrum*, **45**(5), 25–29.

Clarke, R. (1993). Asimov's laws for robotics: Implications for information technology – Part 1. *Computer*, **26**(12), 53–61.

Clarke, R. (1994). Asimov's laws for robotics: Implications for information technology – Part 2. *Computer*, **27**(1), 57–65.

Conrad, K. L., Shiakolas, P. S., and Yih, T. C. (2000). Robotic calibration issues – accuracy, repeatability and calibration. *Proceedings of the 8th Mediterranean Conference on Control & Automation* (MED 2000), July 17–19, 2000, Rio Patras, Greece.

Craig, J. J. (1989). *Introduction to Robotics – Mechanics and Control*, 2nd edn. Reading, MA: Addison-Wesley.

Dagalakis, N. G. (1999). Industrial robotics standards. In *Handbook of Industrial Robotics*, 2nd edn., ed. S. Y. Nof. New York: John Wiley and Sons.

Danielson, P. (1998). Gas loads in vacuum systems. *Vacuum & Thinfilm*, October 1998, 37. Retrieved from the Internet on July 4, 2006 at http://www.vacuumlab.com/articles/.

Danielson, P. (2000). Gas loads and o-rings. *R&D Magazine*, August 2000. Retrieved from the Internet on July 4, 2007 at http://www.vacuumlab.com/articles.

Danielson, P. (2001a). Contamination in vacuum systems, sources and remedies. *R&D Magazine*, May 2001, 45. Retrieved from the Internet on August 28, 2008 at http://www.vacuumlab.com/articles.

Danielson, P. (2001b). Vacuum envelope penetration. *R&D Magazine*, September 2001, 82. Retrieved from the Internet on July 24, 2007 at http://www.vacuumlab.com/articles.

Danielson, P. (2003). Choosing the right vacuum materials. *R&D Magazine*, April 2003. Retrieved from the Internet on August 22, 2006 at http://www.rdmag.com.

Danielson, P. (2004). The use and misuse of O-rings. *R&D Magazine*, April 2004. Retrieved from the Internet on September 23, 2007 at http://www.rdmag.com.

Danielson, P. (2005). Conditioning vacuum chambers. *R&D Magazine*, October 2005. Retrieved from the Internet on September 27, 2007 at http://www.rdmag.com.

Defosse, M. (2004). Automatica: Integrated automation, the next big thing. *Modern Plastics* (online), October 2004. Retrieved from the Internet on March 2, 2005 at http://www.modplas.com.

Denavit, J. and Hartenberg, R. S. (1955). A kinematic notation for lower-pair mechanisms based on matrices. *ASME Journal of Applied Mechanics*, June 1955, 215–221.

Donovan, R. P. (2001). *Contamination-Free Manufacturing for Semiconductors and Other Precision Products*. London: CRC Press.

Egan, G. (2000). The automated wet bench in semiconductor cleanrooms. *Cleanroom Technology*, April 2000.

ESDA – Electrostatic Discharge Association (2001a). *Fundamentals of Electrostatic Discharge – Part 1*. Retrieved from the Internet on March 28, 2008 at http://www.esda.org/esd_fundamentals.html.

ESDA – Electrostatic Discharge Association (2001b). *Fundamentals of Electrostatic Discharge – Part 2*. Retrieved from the Internet on March 27, 2008 at http://www.esda.org.

Featherstone, R. (2007). *Rigid Body Dynamics Algorithms*. New York: Springer.

Ferrotec Corporation (2001). *Ferrofluidic Vacuum Rotary Feedthroughs*, catalog and product specifications. Retrieved from the Internet on June 1, 2007 at http://www.ferrotec.com.

Gao, J. P. G. (2004). Contamination control in photolithography of SAW filter wafers. *Controlled Environments Magazine*, October 2004. Retrieved from the Internet on January 20, 2008 at http://www.cemag.us.

Grecu, H. and Kücher, P. (2000). Strategic importance and value of joint ventures in semiconductor technology development. *Future Fab International*, **8**. Retrieved from the Internet on September 25, 2007 at http://www.future-fab.com.

Halliday, B. S. (1987a). An introduction to materials for use in vacuum. *Vacuum*, **37**(8, 9), 583–585.

Halliday, B. S. (1987b). Cleaning materials and components for vacuum use. *Vacuum*, **37**(8, 9), 587–591.

Hayati, S. A. and Mirmirani, M. (1985). Improving the absolute position accuracy of robot manipulators. *Journal of Robotic Systems*, **2**(4), 397–413.

Helgeland, W. (2002). *Magnetic Fluid Feedthrough Primer.* Rigaku technical report. Retrieved from the Internet on July 4, 2006 at http://www.rigaku.com.

Higuchi, M., Kawamura, T., Kaikogi, T., Murata, T., and Kawaguchi, M. (2003). Cleanroom robots – Clean material handling originated from plant equipment inspection. *Technical Review*, **40**(5), Mitsubishi Heavy Industries.

Hoffman, J. M. (2003). More metal parts go to composites. *Machine Design*, October 2003. Retrieved from the Internet on March 22, 2008 at http://www.machinedesign.com.

Hoffman, J. M. (2008). Stronger and lighter – Composites make their mark. *Machine Design*, March 2008. Retrieved from the Internet on March 22, 2008 at http://www.machinedesign.com.

Hollerbach, J. M. (1989). Kinematics and dynamics for control. In *Robotics Science*, ed. M. Brady. Cambridge, MA: MIT Press.

Hristu-Varsakelis, D. and Levine, W. S., eds. (2005). *Handbook of Networked and Embedded Control Systems*. Boston, MA: Birkhäuser.

Hudgens, J. (1992). *Static Robot Compliance and Metrology Procedures with Applications to a Light Machining Robot*. Unpublished Ph.D. dissertation, University of Texas at Austin.

Hudgens, J., Cox, D., and Tesar, D. (2000). Classification structure and compliance modeling for serial manipulators. *Proceedings of the IEEE International Conference on Robotics and Automation*, April 2000, San Francisco, CA.

ICE – Integrated Circuit Engineering Corporation (1997). *Cost-effective IC Manufacturing 1998–1999*. Report ISBN 1-877750-60-3. Scottsdale, AZ: ICE.

IFR – International Federation of Robotics (2006). *World Robotics 2006 – Executive Summary.* Frankfurt am Main, Germany: IFR.

IFR – International Federation of Robotics (2007). *World Robotics 2007 – Executive Summary.* Frankfurt am Main, Germany: IFR.

IFR – International Federation of Robotics (2008a). *World Robotics 2008 – Welcome to the Press Conference*. Frankfurt am Main, Germany: IFR. Retrieved from the Internet on July 9, 2009 at http://www.worldrobotics.org.

IFR – International Federation of Robotics (2008b). *Industrial Robots are Meeting the Changing Demands of Their Customers*. Frankfurt am Main, Germany: IFR. Retrieved from the Internet on March 10, 2009 at http://www.worldrobotics.org.

Isom, J. (2004). History of robotics. Retrieved from the Internet on February 28, 2008 at http://www.robotics.megagiant.com/history.html.

Jerz, D. G. (2002). R.U.R. (Rossum's Universal Robots). Retrieved from the Internet on February 28, 2008 at http://jerz.setonhill.edu/resources/RUR.

Kahaner, D.K. (1991). *Robots and Use in Japanese Industry.* Report, Asian Technology Infomation Program (ATIP), February 1991. Tokyo: ATIP.

Kanetomo, M., Kashima, H., and Suzuki, T. (1997). Wafer-transfer robot for use in ultrahigh vacuum. *Journal of Vacuum Science and Technology*, **15**(3), Part 2, 1384–1387.

Kraz, V. (2003). Effective management of an ESD environment in production. *Conformity Magazine*, April 2003.

Kunii, I. M. and Port, O. (2001). Robots. *Business Week* (online), March 19, 2001. Retrieved from the Internet on February 28, 2008 at http://www.businessweek.com.

Kuo, B. C. (1981), *Automatic Control Systems*. Englewood Cliffs, NJ: Prentice Hall.

Levit, B., Steinman, A., and Avery, C. S. (2007). Improving particle contamination control with in-tool air ionization. *Solid State Technology*, **50**(3), 48–50.

Ljung, L. (1987). *System Identification*. Englewood Cliffs, NJ: Prentice Hall.

Manji, J. F. (2000). Clean room robots sweep up big savings in semiconductor applications. *Robotics Online*, 2000. Retrieved from the Internet on November 29, 2004 at http://www.roboticsonline.com.

Marcus, M. (2005). *Performance of Hybrid Band-Worm Drives*. Technical report ObservatoryScope, Ellijay, GA. Retrieved from the Internet on August 10, 2007 at http://www.observatoryscope.com.

Marquardt, N. (1999). Introduction to the Principles of Vacuum Physics. Lecture. Geneva: CERN Accelerator School on Vacuum Technology.

Mathia, K. (1996). *Solutions of Linear Equations and a Class of Nonlinear Equations using Recurrent Neural Networks*. Unpublished Ph.D. dissertation, Portland State University.

Mathia, K. and Aalund, M. P. (2002). Full-on for the information age. *European Semiconductor*, **3**, 19–24.

Mathia, K. and Saeks, R. (1994). Inverse kinematics via linear dynamic networks. *Proceedings of the World Congress on Neural Networks 1994* (WCNN'94), **2**, 47–53, San Diego, CA.

Mattox, D. M. (1998). *Handbook of Physical Vapor Deposition (PVD) Processing*. Bracknell, UK: Noyes Publications.

McIntyre, J. R. (1997). *Japan's Technical Standards – Implications for Global Trade and Competitiveness*. London: Quorum Books.

Megahed, S. M. (1993). *Principles of Robot Modeling and Simulation*. Chichester, UK: John Wiley and Sons.

Mohler, R. R. (1991). *Nonlinear Systems, Part 1 – Dynamics and Control*. Englewood Cliffs, NJ: Prentice Hall.

Murvihill, W. (1999). Advances in robotic vacuum. *Vacuum & Thinfilm*, March 1999.

Musser, C. W. (1960). The harmonic drive. *Machine Design*, April 1960, 166–179.

Nakayama, A., Kimura, H., Watanabe, K., et al. (2005). Low outgassing cables for clean room. *Hitachi Cable Review*, **24**.

Nof, S. Y. (1999). *Handbook of Industrial Robotics*, 2nd edn. New York: John Wiley and Sons.

Ogata, K. (1990). *Modern Control Engineering*, 2nd edn. Englewood Cliffs, NJ: Prentice Hall.

O'Halloran, M., Grout, R., and Pettengill, N. (2002). Oversized cleanrooms assume huge role in emerging flat-panel display production. *Cleanrooms*, August 2002.

Oppenheim, A. V. and Schafer, R. W. (1989). *Discrete-time Signal Processing*. Englewood Cliffs, NJ: Prentice Hall.

Papoulis, A. (1991). *Probability, Random Variables, and Stochastic Processes*, 3rd edn. New York: McGraw-Hill.

Penrose, R. (1955). A generalized inverse for matrices. *Proceedings of the Cambridge Philosophical Society*, **51**, 406–413.

Pentagon Technologies (2001). *Particle Profiling of Cable Carriers*. Test report, March 2001. Fremont, CA.

Pentagon Technologies (2002). *Particle Profile of the IRSI Robot.* Test report. Livermore, CA.

Pentagon Technologies (2004). *Particle Profile of the IRSI Robot* – Addendum. Test report. Livermore, CA.

Pfeiffer, O., Ayre, A., and Keydel, C. (2003). *Embedded Networking with CAN and CANopen*. San Clemente, CA: RTC Books.

Rigaku Corporation (2005). *Magnetic Fluid Feedthrough Primer.* Technical report. Retrieved from the Internet on September 27, 2009 at http://www.rigaku.com/vacuum/primer.html.

Saravanan, C., Liu, Z., Yang, W., Swisher, M. F., and Tang, A. (2004). Investigating the impact of backside defect inspection on process development and yields. *MICRO Magazine*, April 2004.

Sasaki, Y. T. (2007). Reducing SS 304/316 hydrogen outgassing to $2 \cdot 10^{-15}$ (torr l cm^{-2} s^{-1}). *Journal of Vacuum Science and Technology: Vacuum, Surfaces, and Films*, **25**(4), 1309–1311.

Schenker, P. S., Das, H., and Ohm T. R. (1995). A new robot for high dexterity microsurgery. *Proceedings of the Conference on Computer Vision, Virtual Reality, and Robotics in Medicine*, April 3–6, 1995, Nice, France.

Schicht, H. (2003). The ISO contamination control standards. *European Journal of Parenteral Sciences*, **8**(2), 37–42.

Schroth, T. and Caesar, T. (2001). EN 1822: The standard that greatly impacted the European cleanrooms market. *Cleanrooms*, October 2001.

Schubert, S. (2005). Taking robots for a ride. *CNN Money*, August 1, 2005. Retrieved from the Internet on February 26, 2008 at http://www.money.cnn.com/magazines.

Sciavicco, L. and Siciliano, B. (1996). *Modeling and Control of Robot Manipulators*. New York: McGraw-Hill.

Shiakolas, P. S., Conrad, K. L., and Yih, T. C. (2002). On the accuracy, repeatability, and degree of influence of kinematics parameters for industrial robots. *International Journal of Modeling and Simulation*, **22**(3), 245–254.

Sontag, E. D. (1990). *Mathematical Control Theory.* New York: Springer Verlag.

Sparrow, N. (2004). Virtual leak diameter standard provides alternative to conventional leak testing. *European Medical Device Manufacturer*, October 2004.

Spong, M. and Vidyasagar, M. (1989). *Robot Dynamics and Control.* New York: John Wiley and Sons.

Stoianovici, D., Cleary, K., Patriciu, A., et al. (2003). AcuBot: A robot for radiological interventions. *IEEE Transactions on Robotics and Automation*, **19**(5), 927–930.

Strong, J. (1938). *Procedures in Experimental Physics*. Bradley, IL: Lindsay Publications.

Sutera, S. P. and Skalak, R. (1993). The history of Poiseuille's law. *Annual Review of Fluid Mechanics*, **25**, 1–19.

Thomson, W. (1996). *Theory of Vibration With Applications*, 4th edn. London: Taylor & Francis.

Thong, Y. K., Woolfson, M. S., Crowe, J. A., Hayes-Gill, B. R., and Jones, D. A. (2004). Numerical double integration of acceleration measurements in noise. *Science Direct*, **36**(1), 73–92.

Thornton, J. (2002). Robot safety: ANSI/RIA R15.06-1999 and savvy safeguarding for robotic workcells. *Robotics Online*, 2002. Retrieved from the Internet on December 2, 2004 at http://www.roboticsonline.com.

Tolani, D., Goswami, A., and Badler N. I. (2000). Real-time inverse kinematics techniques for anthropomorphic limbs. *Graphical Models*, **62**, 353–388.

Tolliver, D. L., ed. (1988). *Handbook of Contamination Control in Microelectronics: Principles, Applications and Technology.* Westwood, NJ: Noyes Publications.

Urban, R. and Hamid, S. (2005). Extreme-environment lubricant maintains extrusion cleanroom. *Plastics Machinery and Auxiliaries*, March/April 2005. Retrieved from the Internet on August 15, 2007 at http://www.pma-magazine.com.

Vogler, D. (2002). Evolving business strategies are key to an upbeat 300 mm outlook. *Solid State Technology*, October 2002.

Watanabe, F. (2001). Mechanism of ultralow outgassing rates in pure copper and chromium–copper alloy vacuum chambers: Reexamination by the pressure-rise method. *Journal of Vacuum Science & Technology: Vacuum, Surfaces, and Films*, **19**(2), 640–645.

Westerberg, L. (1999). *The Best Laid Schemes…*. Lecture notes. Geneva: CERN Accelerator School on Vacuum Technology.

White, F. M. (2003). *Fluid Mechanics*. McGraw-Hill Series in Mechanical Engineering. New York: McGraw-Hill.

Winters, P. (2005). IEST updates and improves its Recommended Practice for testing ULPA filters. *Cleanrooms*, June 2005.

Younkin, G. W. (1996). *Industrial Servo Control Systems*. New York: Dekker.

Zhuang, H. and Roth, Z. S. (1996). *Camera-aided Robot Calibration*. Boca Raton, FL: CRC Press.

Index